Connectionist Models of Cognition and Perception II

PROGRESS IN NEURAL PROCESSING*

Series Advisor

Alan Murray *(University of Edinburgh)*

Vol. 3: Parallel Implementation of Backpropagation Neural Networks on Transputers: A Study of Training Set Parallelism
by P. Saratchandran, N. Sundararajan & S.-K. Foo

Vol. 4: Analogue Imprecision in MLP Training
by Peter J. Edwards & Alan F. Murray

Vol. 5: Applications of Neural Networks in Environment, Energy, and Health
Eds. Paul E. Keller, Sherif Hashem, Lars J. Kangas & Richard T. Kouzes

Vol. 6: Neural Modeling of Brain and Cognitive Disorders
Eds. James A. Reggia, Eytan Ruppin & Rita Sloan Berndt

Vol. 7: Decision Technologies for Financial Engineering
Eds. Andreas S. Weigend, Yaser Abu-Mostafa & A.-Paul N. Refenes

Vol. 8: Neural Networks: Best Practice in Europe
Eds. Bert Kappen & Stan Gielen

Vol. 9: RAM-Based Neural Networks
Ed. James Austin

Vol. 10: Neuromorphic Systems: Engineering Silicon from Neurobiology
Eds. Leslie S. Smith & Alister Hamilton

Vol. 11: Radial Basis Function Neural Networks with Sequential Learning
Eds. N. Sundararajan, P. Saratchandran & Y.-W. Lu

Vol. 12: Disorder Versus Order in Brain Function: Essays in Theoretical Neurobiology
Eds. P. Århem, C. Blomberg & H. Liljenström

Vol. 13: Business Applications of Neural Networks: The State-of-the-Art of Real-World Applications
Eds. Paulo J. G. Lisboa, Bill Edisbury & Alfredo Vellido

Vol. 14: Connectionist Models of Cognition and Perception
Eds. John A. Bullinaria & Will Lowe

*For the complete list of titles in this series, please write to the Publisher.

Progress in Neural Processing • 15

Proceedings of the Eighth Neural Computation and Psychology Workshop

Connectionist Models of Cognition and Perception II

University of Kent, UK 28 – 30 August 2003

Editors

Howard Bowman
University of Kent, UK

Christophe Labiouse
University of Liege, Belgium

World Scientific

NEW JERSEY • LONDON • SINGAPORE • SHANGHAI • HONG KONG • TAIPEI • CHENNAI

Published by

World Scientific Publishing Co. Pte. Ltd.
5 Toh Tuck Link, Singapore 596224
USA office: Suite 202, 1060 Main Street, River Edge, NJ 07661
UK office: 57 Shelton Street, Covent Garden, London WC2H 9HE

British Library Cataloguing-in-Publication Data
A catalogue record for this book is available from the British Library.

CONNECTIONIST MODELS OF COGNITION, PERCEPTION AND EMOTION
Proceedings of the Eighth Neural Computation and Psychology Workshop

ISBN 981-238-805-2

Printed in Singapore by World Scientific Printers (S) Pte Ltd

Preface

This volume collects together refereed versions of papers presented at the Eighth Neural Computation and Psychology Workshop (NCPW 8), which was held at the University of Kent at Canterbury, England in August 2003. The NCPW series is a well-established and lively forum that brings together researchers from such diverse disciplines as artificial intelligence, cognitive science, computer science, neuroscience, philosophy and psychology. Thirty-five papers were presented at the event, of which eight were invited papers. More than fifty participants attended the workshop that drew researchers from England, Ireland, the Netherlands, Belgium, France, Israel, Spain, Finland, Germany, and the United States. This large and cosmopolitan audience is evidence of the ever-increasing importance of neural network modelling in the cognitive sciences at the dawn of the twenty-first century. The overarching theme of this eighth workshop in the series was Connectionist Models of Cognition and Perception.

The structure of this book broadly follows the session structure of the event, although some papers have been repositioned in order to increase the coherence of the publication. The book is divided into nine sections, which cross the spectrum of cognitive phenomena and reflect the extent of research being undertaken on connectionist modelling of cognition. Specifically, the book contains sections on Memory, Vision, Action and Navigation, Developmental Processes, Category Acquisition, Attention, High Level Cognition and Implementation Issues, Language and Speech, and Cognitive Architectures and Binding.

The Memory section contains two papers. Davelaar and Usher present a model of active maintenance in working memory and their computational model sheds light on the distinction between short-term and long-term memories. Musca, Rousset and Ans investigate the effect of material structure on retroactive and proactive interference. They propose a dual-network architecture to account for these effects in association learning. The Vision section is also comprised of two papers. Firstly, Hurri, Vayrnynen and Hyarinen consider the spatio-temporal properties of simple-cells in V1 based on their temporal coherence and with the help of independent component analysis. Then Karanka and Luque explore how a simple-recurrent network can be used to predict time-to-collision.

The next section is devoted to Action and Navigation. Of the four papers in this section, the first (by Theofilou, Destrebecqz and Cleeremans) uses forward models to explore human's ability to learn sequences. The second paper in the section, which is by Richardson, Davey, Peters and Done, considers how human character production can be modelled using recurrent neural networks. Then, Girard, Filliat, Meyer, Berthoz and Guillot explore basal ganglia-based control architectures, which enable robots to select actions and navigate. The final paper in this section (by Schenck and Moller) presents a system that can learn saccadic eye movements in a staged manner.

Developmental Processes are considered in the next section, which comprises three papers. The first paper is by Westermann and Mareschal who provide a RBF-like neural network account for an observed asymmetry that occurs in the categorization of cats and dogs by 3-4 month old children. Then Carlson and Triesch use reinforcement learning to provide a nurture-based explanation of the emergence of gaze following during infant development. Finally, Levy's paper, which concludes this section, reviews the ways in which connectionist models have been used to explain autism.

Next we come to a section focused on Category Acquisition. The initial paper in this section is by Joyce and Cottrell and it provides a connectionist explanation for why and how a brain area specifically dedicated to expert discrimination may have developed. Then Fink, Ben-Shakhar and Horn use a neural network to contrast the role of two factors that govern feature creation: the informative value and the degree of parsimony of the feature set. Finally, Mermillod, Guyader and Chauvin answer the question of whether the energy spectrum of Gabor wavelets can represent sufficient information for recognition and classification tasks.

The next section is devoted to neural network models of attentional processes. Taylor approaches the issue of consciousness through the provision of an engineering control account of attention and motor control. Then Heinke, Humphreys and Tweed present an extension of the Selective Attention for Identification network, which models visual search. Next, Bowman, Wyble and Barnard, present a neural network model of the deployment of attention over time, in the context of the attentional blink. The final paper of the section is by Bartos and it adds an attentional mechanism to the configural-cue model of stimulus representation.

High Level Cognition and Implementation Issues are considered in the next section, which begins with a paper by Leech, Mareschal and Cooper on the application of attractor networks to analogical reasoning. Then Bullinaria discusses a number of simulations that consider how irrational behaviour could emerge from evolution. Next, Van Overwalle explores how connectionism can be applied in the social psychological context of the multiple inference model of person impression formation. In the final paper in this section Connolly, Marian and Reilly present several algorithms for the simulation of spiking neural networks on single processor systems.

The penultimate section addresses Language and Speech, with the first paper (by Shillcock and Monaghan) considering visual word recognition using a split-fovea model. In the next paper, Hayes, Murphy, Davey and Smith use simple recurrent networks to provide a nurture-based explanation of the formation of English noun-noun plurals. Then Moscoso del Prado Martin, Schreuder and Baayen consider how to build distributed representations of word forms by accumulation of expectations. Finally, Hammerton compares connectionist models of speech segmentation in the context of the utterance boundary strategy.

The last section of the book focuses on Cognitive Architectures and Binding. Firstly, Borisyuk and Kazanovich present an oscillation-based cognitive model of brain function and then Mair, Shepperd, Cartwright, Kirsopp, Premraj and Heathcote present experimental findings on object feature binding and then discuss how these findings could be implemented in a neural network.

We would like to thank all those that attended NCPW'8 and made the event such a stimulating occasion. We would particularly like to thank our eight invited speakers: John Bullinaria, Gary Cottrell, Bob French, Peter Hancock, Richard Shillcock, Chris Solomon, John Taylor and Marius Usher. All of whom gave thought-provoking talks which fully reflected the state of the art of research in their chosen area. We would especially like to pay thanks to Gary Cottrell, who despite heavy jet lag, enthusiastically contributed throughout the event, both intellectually and socially.

We would also like to thank the following for reviewing papers for the proceedings: Paul Bartos, Roman Borisyuk, John Bullinaria, Axel

Cleeremans, Gary Cottrell, Eddy Davelaar, Michael Fink, Bob French, Benoit Girard, Peter Hancock, Dietmar Heinke, Jarmo Hurri, Robert Leech, Martin Le Voi, Joe Levy, Fermin Moscoso del Prado Martin, Martial Mermillod, Serban Musca, Ronan Reilly, Corina Sas, Wolfram Schenk, Richard Shillcock, Dionyssios Theofilou, Jochen Triesch, Marius Usher, Tim Valentine, Frank van Overwalle and Brad Wyble.

On the organisational side, we would like to pay special thanks to Jenny Oatley and Deborah Sowrey who provided excellent secretarial support before, after and during the event. In addition, Colin Johnson, Miguel Mendao, Vikki Roberts and Brad Wyble freely gave of their time in order to provide organisational support for the event. Finally, we wish to thank the Computing Laboratory at the University of Kent at Canterbury, which provided considerable financial support for the event.

Howard Bowman & Christophe Labiouse[1]
December 2003

[1] Christophe Labiouse is a Research Fellow of the Belgian National Fund of Scientific Research (FNRS).

Contents

Developmental Processes

Category Acquisition

Attention

High Level Cognition and Implementation Issues

Language and Speech

Cognitive Architectures and Binding

Memory

AN EXTENDED BUFFER MODEL FOR ACTIVE MAINTENANCE AND SELECTIVE UPDATING

EDDY J. DAVELAAR AND MARIUS USHER

School of Psychology, Birkbeck College, University of London, Malet street, London, WC1E 7HX, United Kingdom

In previous work, we developed a neurocomputational model of list memory, based on neural mechanisms, such as recurrent self-excitation and global inhibition that implement a short-term memory activation-buffer. Here, we compare this activation-buffer with a series of mathematical buffer models that originate from the 1960s, with special emphasis on presentation rate effects. We then propose an extension of the activation-buffer to address the process of selectively updating the buffer contents, which is critical for modeling working memory and complex higher-level cognition.

1. Introduction

Many models of human memory have been developed in psychology since the early 1960s[1,2] addressing a variety of tasks such as the immediate free recall. Most of these models were abstract-mathematical (rather than neuro-computational models) and their advantage is being simple and transparent, thus easy to understand. Recently, a shift towards neurocomputational models is taking place[3-9], which due to their increased complexity can account for a wider range of data including the effects of neuropsychological dissociations[6,10]. Nevertheless such models are more complicated and therefore more difficult to understand.

Here we start (section 2) by comparing our previous neurocomputational model of active memory with a series of buffer models, suggesting a way to reduce it (or extend the buffer models) so as to capture some important data in immediate free recall. In section 3, we propose ways in which our activation-buffer could be extended in order to address working memory processes, such as selective updating.

2. Mathematical and Neurocomputational Buffer Models

In the field of memory research, the free recall paradigm has led to many theoretical viewpoints and debates. In the immediate free recall paradigm, participants are required to report, in any order, as many items from a list that has been presented to them. The typical result is better recall performance for items that were presented at the beginning and at the end of the list, the primacy and (S-shaped) recency effect[11], respectively. One view of the recency effect is that the end-of-list items still reside in a limited-capacity short-term buffer from which they are reported without error[1,2,4-6,12]. In this section, we compare mathematical buffer models, which have been used in early psychological theories to explain free recall performance, with our neurocomputational activation-buffer, with a special emphasis on the effect of presentation rate.

The models are compared on four measures. First, the serial position functions present the probability that an item is in the buffer at the end of a sequence of twelve items (1st column in Figures 1 and 4). Second, we compare the distributions of the number of items in the buffer at the end of the sequence (2nd column in Figures 1 and 4). Third, we compare the probability that a new item will enter the buffer as a function of the presentation rate and the number of items already in the buffer (3rd column in Figures 1 and 4). This comparison will turn out to provide valuable information related to the effect of presentation rate. The fourth and last measure on which the buffer models are compared is the distribution of probabilities that an item will be displaced from the buffer as a function of the number of items already in the buffer and the relative recency of the displaced item (4th column in Figures 1 and 4).

2.1. *Mathematical Buffer Models*

Three mathematical buffer models that have been used in the psychological literature are the *random-buffer*[1,2], the *knock-out buffer*[1,13] and the *variable knock-out buffer*[14]. Due to space-limitations a thorough analysis including other buffer models will be left for a future project.

Random-buffer model (RB)

The first buffer model is that in which the buffer consists of a fixed number of slots, r. When the buffer is full to capacity, a displacement process randomly (and with equal probability) selects which of the r slots will be emptied and be occupied with the newly presented item. The top row of Figure 1 shows the results of such a model. The left panel shows the serial position function for a buffer with capacity 3 and with capacity 4. These are exponential functions with base (r-1)/r and exponent -(sp+1), where sp indicates the recency of the item (-1 being the most recent). The second panel shows the distribution of the effective capacity at the end of a sequence of twelve items. As this is a fixed-capacity buffer, the distribution is centred on r. The third panel shows the probability that a presented item will enter the buffer as a function of presentation duration and number of items already in the buffer. By definition, all the mathematical buffer models described here have a probability of unity that an item enters the buffer, regardless of presentation duration and current buffer contents. The right-most panel shows the distribution of the probabilities that a buffer-item will be displaced from the buffer, as a function of its relative recency (the item that has been in the buffer the longest will have a relative recency of –r, whereas the latest addition to the buffer contents has a relative recency of –1). No surprise here that with random displacement, the distribution is uniform with probability 1/r.

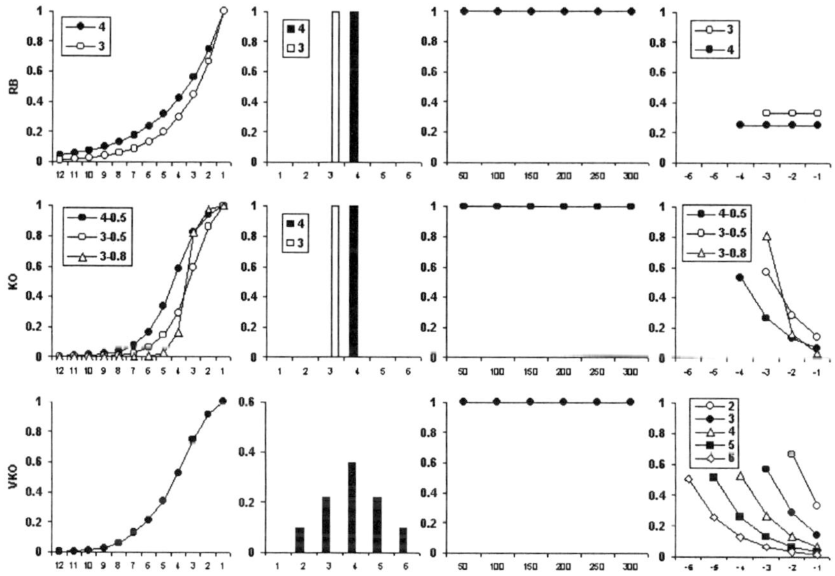

Figure 1. Comparisons of three mathematical buffer models. From top to bottom, results are presented for the random-buffer (RB), the knock-out buffer (KO) and the variable knock-out buffer (VKO). The results show serial position functions at different levels of capacity and displacement parameters (1st column), distribution of the number of items in the buffer after a twelve items sequence (2nd column), probability of a new item entering the buffer as a function of presentation duration (see abscissa) and number of items already in the buffer (3rd column) and the distribution of displacement probabilities as a function of the number of items in the buffer and the relative recency (4th column).

Knock-out buffer model (KO)

A variant of the random-buffer model is one in which the displacement process is such that the probability that an item is displaced from the buffer increases with the duration that the item has been in the buffer. This has been referred to as the knock-out buffer model[1,13]. The probability, d_i, of displacing item i, depends on the capacity r and a parameter δ that governs the slope of the displacement distribution (see Equation 1). The second row of Figure 1 shows the results. In the first panel, serial positions are presented for the model with capacity 3 and 4 and with δ=0.5 and δ=0.8. What is immediately apparent is that all serial position functions are S-shaped and that this increases with δ (compare 3-0.5 with 3-0.8). The distributions of the capacity (second panel) and the probability of entering the buffer (third panel) are the same as the random-buffer model. The right-most panel presents the distribution of displacement probabilities, which is a clear departure from the random process.

$$d_i = \delta(1-\delta)^{i-1}/[1-(1-\delta)^r] \qquad \text{(Eq. 1)}$$

Variable knock-out buffer model (VKO)

The third buffer model is one that extends the knock-out buffer and in which for every trial in a simulation the capacity r is drawn from a distribution of capacities[14]. This has the benefit of allowing more flexibility, as a participant's effective capacity may also depend on internal fluctuations in attention. The third row in Figure 1 shows the results of the variable-knock-out buffer model. The serial position function is basically a weighted aggregate of the various knock-out buffers within it. The second panel shows the distribution of capacities from which r was drawn. As an item always enters the buffer (third panel), the distribution of the displacement probabilities (fourth panel) is a collection of distributions at various capacities (which are all at unity).

2.2. *Neurocomputational Activation-Buffer Model*

We[4-6] developed a neurocomputational model of immediate free recall that is formulated within the Hebbian framework[15,16] with short-term memory (or primary memory[16]) being mediated by the current set of activated neuronal representations and long-term memory (or secondary memory[16]) being mediated by the connections between the activated subset and an episodic contextual system.

Figure 2 presents the architecture of our neurocomputational model. Each unit represents an assembly of interconnected neurons. When a stimulus is presented to the system, the corresponding representation will receive sensory input and increases in activation. To simplify, we use a single parameter, α, for the self-excitation. The self-excitation enables the unit to remain active above threshold after the sensory input has been taken away. Within the system, there is a global competition. This can be considered as originating from a general pool of inhibitory inter-neurons and has the effect of limiting the number of representations that can be active simultaneously.

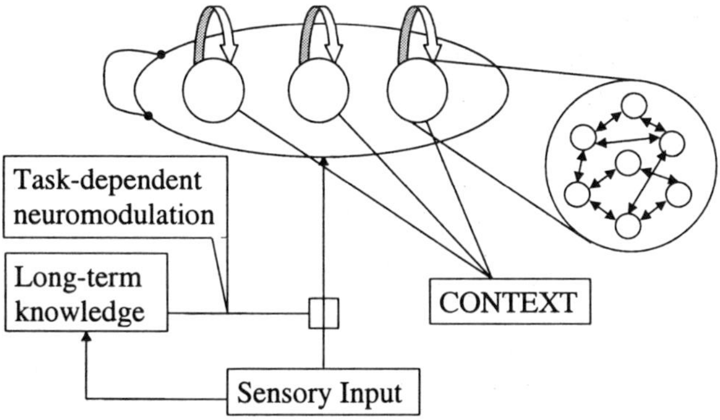

Figure 2. Architecture of our neurocomputational approach to list memory. The ellipse represents the activation-buffer, with units representing cell-assemblies and is addressed in section 2. The arrow ending in closed circles denotes global inhibition. The units form episodic links with a context representation. Sensory input goes directly into the activation-buffer after being neuro-modulated. The specific architecture of how the neuromodulation is driven is arbitrarily chosen and does not change the discussion on selective updating in section 3.

Our model is used in real-time, where all units are updated at every time-step according to a leaky integrator differential equation of which Equation 2 is the numerical solution in discrete time steps.

$$x_i(t+1) = \lambda x_i(t) + [1-\lambda][\alpha F(x_i(t)) - \beta \Sigma F(x_j(t)) + I_i(t) + \xi] \qquad \text{(Eq. 2)}$$

Here, λ=0.98 is the decay constant, α=2.0 the self-excitation, β=0.15 the global inhibition, I(t)=0.33 (0, when no input is presented) the sensory input at time t and ξ the zero-mean Gaussian noise with standard deviation σ=0.5. F(x) is the output activation function[8] MAX[0, x/(1+x)], which is similar to the threshold-linear function with the addition of a saturating non-linearity.

We also assume that units that are activated above threshold are encoded in episodic memory, which comprises of a matrix of Hebbian connection weights between the items and a context system. However, here we focus primarily on the dynamics of the activation-buffer, which are illustrated in Figure 3. Twelve units are sequentially presented with sensory input for 250 (left panel) or 100 (right panel) iterations corresponding to a typical experimental procedure where a list of twelve items are presented sequentially on a computer screen at different presentation rates. Each line corresponds to the output activation, $F(x_i)$, of a given unit i. The left panel of Figure 3 shows the set of activation trajectories when the presentation rate is relatively slow and the right panel shows the set of trajectories at fast presentation rate. Two aspects can be observed. First, units remain active after stimulus offset, which is due to the self-recurrent excitation.

Second, several units can be active simultaneously and there is an upper limit to the number of units that are active above threshold, which reflects the capacity limitation due to global inhibition.

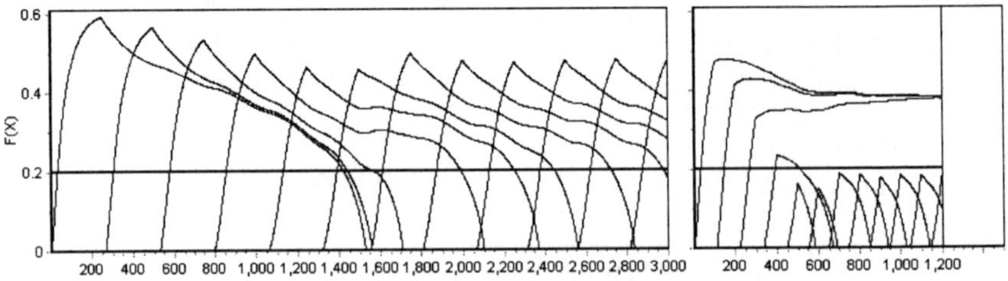

Figure 3. Activation trajectories of twelve sequentially activated units at slow (left) and fast (right) presentation rates. Time-steps are set along the abscissa and the output activation on the ordinate. The horizontal line [at F(X)=0.2] represents the activation threshold above which an item is said to be in the buffer.

With increase in the presentation rate, the activation-buffer changes its behaviour. First, the units reach a lower level of activation compared to the condition with slow presentation rate. This merely reflects the limited time that is given for the units to accrue. Second, with the same structural parameters, the number of active units is smaller than the number of active units at slow presentation rates, implying that the effective capacity of the activation-buffer depends on external variables like presentation rate. Elsewhere[4], we have shown that the system will not exceed a certain upper limit given a wide range of presentation rates[a]. Third, whereas at slow presentation rates the unit to be displaced (de-activated) from the current buffer contents, is typically one that has been in the buffer (above threshold) the longest, at fast presentation rates the buffer only maintains the first few items and blocks out any subsequently presented item. In other words, at low presentation rates, the activation-buffer is a limited-capacity buffer system with a knock-out displacement process, while at fast presentation rates the probability of entering the buffer is greatly diminished. This prediction is fully due to the limited time available that a unit can be activated to the extent that it can overcome the amount of inhibition already in the system. As the first item enters an empty buffer, it will not have to overcome this sort of inhibition, giving it an advantage over subsequent activated units.

2.3. *The Effect of Presentation Rate on Buffer Dynamics*

Figure 3 shows activation trajectories for the activation-buffer under slow and fast presentation rates. In the top row of Figure 4 the results of the activation-

---

[a] In fact, the effective capacity shows an inverted U-curve with presentation rate.

buffer are presented on the same four measures we examined for the three mathematical buffer models, so that it can be compared with them (cf. Figure 1). First of all, the left panel shows the serial position functions for three presentation rates (here presented as durations). With slow presentation rates (250 iterations per item), the serial position function is recency-biased and S-shaped, whereas with intermediate rates (150 iterations per item), the function is recency-biased, J-shaped (exponentially-shaped) and some primacy items are maintained. However, with fast presentation rates (50 iterations per item), the serial position function is primacy-biased and J-shaped. It is important to remember that the serial position functions represent the probability of items presented at that position in the sequence still being active above threshold. No Hebbian weight-changes or other long-term memory processes are incorporated. This switch from recency to primacy with increase in presentation rate was verified in an experiment[6]. The activation-buffer maintains less items under fast than under slow presentation rates, as indicated by the shift in the distribution of the number of active items at the end of a twelve-item sequence. As mentioned before, in this range of presentation rates, the effective capacity is negatively correlated with the presentation rate.

Two major differences between the activation-buffer and the mathematical buffers were observed. First, for the activation-buffer, the probability that an item will enter the buffer depends on the presentation rate and the number of items already in the buffer. In the activation-buffer, increasing the presentation rate decreases the probability that a unit can be activated to such a level at which it can overcome the inhibition in the system, which increases with the number of items already in the buffer. This dual-relationship leads to the complex interaction depicted in the third panel. Second, with slow presentation rates the distribution of displacement probabilities for the activation-buffer suggest a knock-out displacement process (see fourth panel). With fast presentation rates the distribution becomes more flat (not shown). This suggests a displacement rule that is rate-dependent, such that with faster presentation rates δ decreases. We focus on extending the knock-out buffer with the rate-dependent probabilities that a presented item will enter the buffer and decrease δ for fast presentation rates.

2.4. *Extending the Knock-Out Buffer*

The above comparisons seem to suggest that the main reason why the mathematical buffers do not predict the shift from a recency-biased to a primacy-biased serial position function with increase in presentation rate is that in those models an item *always* enters the buffer. Although in the original Atkinson and

Shiffrin[1] buffer model, a parameter was included that governed the probability of entering the buffer, simulations estimated its value at around unity, which is consistent with the activation-buffer at slow presentation rates. Re-introducing the parameter and making it dependent on the number of items already in the buffer and presentation rate would allow the mathematical model to accommodate the recency-to-primacy shift.

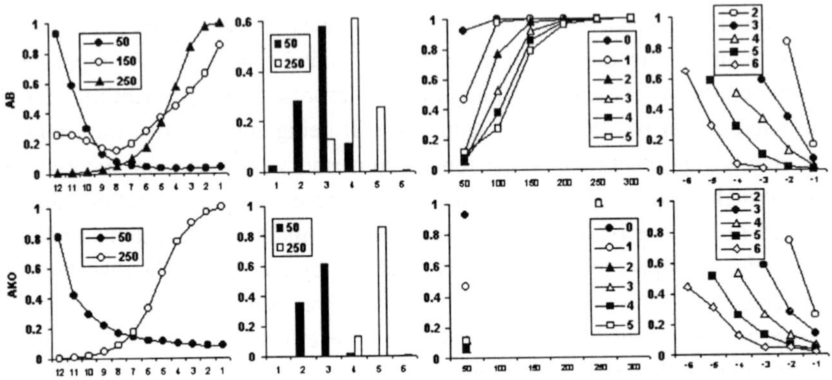

Figure 4. Results for the activation-buffer (AB; top row) and the activation knock-out buffer (AKO; bottom row) on the four measures for slow (250 iterations per item) and fast (50 iterations per item) presentation rates. For the activation-buffer, an intermediate presentation is also shown, indicating a gradual transition from recency-to-primacy bias. Note that for the activation knock-out buffer, the probability of entering the buffer at the two presentation rates are taken directly from those of the activation-buffer. The distribution of displacement probabilities as a function of relative recency is only shown for the slow presentation rate. In the AKO buffer, δ_{slow}=0.5 and δ_{fast}=0.01.

To test this assumption, we added a parameter to the knock-out buffer. We chose to extend the knock-out buffer as it contains the right kind of assumptions that lead to S-shaped serial position functions. Although we used the probabilities obtained with the activation-buffer, we did notice that the relationship between the probability of entering the buffer, the presentation rate and the current capacity can be approximated with a single sigmoidal function. Here, we are only interested in whether adding the probabilities will produce the two main predictions from the activation-buffer. As can be seen in the bottom row of Figure 4, adding the probabilities allows the model to predict the recency-to-primacy shift (first panel) and the decrease in effective capacity with increase in presentation rate (second panel). The rightmost panel shows the distributions of displacement probabilities, which are similar to those of the variable knock-out (third row, Figure 1) and activation-buffer (top row, Figure 4). It is important to realise that the variability in the effective capacity is a consequence of the probabilities that a newly presented item enters the system and the probabilities that an item is displaced from the system.

This exercise suggests that the initial conception of the knock-out buffer with the additional "entry-parameter" by Atkinson and Shiffrin[1] contained the relevant assumptions to predict the recency-to-primacy shift. These assumptions in turn follow naturally from the complex dynamics of the activation-buffer. To summarise, the activation-buffer shows that δ and the entry-probability are inversely related to the presentation rate.

3. Selective Updating of the Buffer

The neurocomputational activation-buffer captures the complex dynamics of short-term memory that are needed to explain the data found in immediate free recall. Within this neurocomputational level of description, it is possible to model the dynamical process of updating the contents of the buffer in accordance with a given task set, as needed to account for cognitive control and working memory[3,7]. The updating task we examine here is one in which a sequence of concrete and abstract nouns is presented with the instructions to remember only those words that represent small things[17]. For example, in the sequence *car, desk, idea, key, plane, staple, giraffe* only the words *key* and *staple* need to be reported. In this example, it is not until *key* is presented that one knows that *car* and *desk* belong to the category of large things and the contents of the buffer is to be updated. However, when *plane* is presented it is already apparent that this belongs to the large-things category and will not even enter the buffer.

As in previous work[3,18], we assume that neuromodulation of sensory input introduces sufficient flexibility to support task-dependent selective updating. In Figure 2 the architecture illustrates a configuration that could lead to task-dependent neuromodulatory control. Sensory input enters the activation-buffer and activates long-term knowledge about the presented item, such as magnitude. With the instruction that small things need to be maintained, words representing small things will provide larger modulated input to the buffer than words representing large things or abstract nouns. In order to capture the essence of the neuromodulation, we represented a sequence of concrete nouns as a sequence of items that vary in the amount of input (I_{target}=0.33, $I_{non\text{-}target}$=0.21). A more detailed model of selective updating with an actual implemented neuromodulatory system is due to space limitations left for future work.

In the left-hand side of Figure 5 the activation trajectories are shown for a sequence of twelve nouns in which nouns 4, 5 and 6 are target nouns and all others are non-targets. As can be seen, the model maintains the first three non-targets until the three targets are presented. After the three targets, none of the non-targets displace the target items: the system has updated the current contents and maintains the targets in the face of distractors. This is due to the targets receiving sufficient effective input to overcome the inhibition driven by the initial non-targets, whereas the non-targets presented after the targets do not receive enough effective input to overcome the inhibition that is then driven by the targets.

In our work on free recall memory[4-6], we assumed that in addition to maintenance in the activation buffer, Hebbian connections are formed between items that are active above threshold and an episodic context system. The strength of these connections is proportional to the integral under the activation trajectories and the threshold. At retrieval, participants can report items from the buffer or trigger a slower competitive retrieval process that uses the episodic Hebbian connections. In Figure 5, it can be seen that non-targets presented before the targets will have stronger episodic connections than the many non-targets presented after the targets, which could lead to more intrusions of non-targets presented before the targets than of those presented after the targets, as reported by Palladino and co-workers[17] (right-hand side of Figure 5).

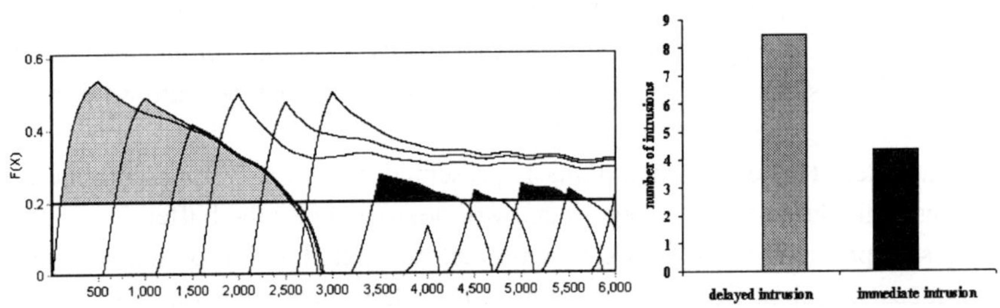

Figure 5. Left: Activation trajectories for a sequence of twelve items in which items 4, 5, and 6 are target items. Note that the targets displace the preceding non-targets and that non of the subsequent non-targets are maintained. The shaded areas correspond to the episodic strengths for the non-targets presented before (grey) and after (black) the target items. Right: Results from Palladino and coworkers on the number of non-target intrusions. Delayed intrusions are before-target non-targets and immediate intrusions are after-target non-targets.

4. Conclusion

In this paper, we compared our neurocomputational activation-buffer with a series of mathematical buffers used in the earlier literature. We found that these buffer models were lacking the flexibility needed to enable them to predict presentation rate effects and we proposed an extension of the knock-out buffer, which may be seen as a reduction of our activation model. We suggest that this illustrates how starting from neurocomputational principles (before reducing to an abstract model) may be productive in modeling psychological processes, since it can ground relatively arbitrary assumptions (in this case the buffer properties). For example, the buffer properties and its capacity limitation follow from mechanisms of recurrent self-excitation (interconnectivity of neurons within an assembly) and the global inhibition (originating from a pool of interneurons). This balance is dynamic and leads to a distribution of the capacities instead of a single capacity value and is affected by external factors like presentation rate, leading to the recency-to-primacy shift.

We have also presented a conceptual extension to the activation-buffer that addresses processes, such as selective updating of the buffer contents. Recently, we[10] showed how the model can account for deviant serial position functions found with neuropsychological patients. We believe that a neurocomputational approach to (short-term) memory not only allows a way to understand how neural principles underlie cognitive behaviour, but also provides a promising platform on which natural extensions can allow for more complex higher-level cognitive processes to be addressed.

Acknowledgments

This work is supported by the Economic and Social Research Council (T026271312). Send correspondence to e.davelaar@psychology.bbk.ac.uk or m.usher@bbk.ac.uk.

References

1. Atkinson, R. C., & Shiffrin, R. M. (1968). Human memory: a proposed system and its control processes. In K. W. Spence & J. T. Spence (Eds.), *The psychology of learning and motivation, Vol 2.* New York: Academic Press.
2. Raaijmakers, J. G. W., & Shiffrin, R. M. (1980). SAM: a theory of probabilistic search of associative memory. In G. Bower (Ed.), *The psychology of learning and motivation, Vol 14.* New York: Academic Press.
3. Braver, T. S., & Cohen, J. D. (2000). On the control of control: the role of dopamine in regulating prefrontal function and working memory. In S. Monsell & J. Driver (Eds.), *Attention and Performance XVIII. Control of cognitive processes.* Cambridge, MA: MIT Press.
4. Davelaar, E. J. (2003). *Active memory: its nature and its role in mnemonic and linguistic behaviour.* Unpublished doctoral dissertation. London: Universtiy of London.
5. Davelaar, E. J., & Usher, M. (2002). An activation-based theory of immediate item memory. In J. A. Bullinaria & W. Lowe (Eds.), *Connectionist models of cognition and perception: proceedings of the seventh neural computation and psychology workshop.* Singapore: World Scientific.

6. Davelaar, E. J., Goshen-Gottstein, Y., Ashkenazi, A., & Usher, M. (submitted). A context/activation model of list memory: dissociating short-term from long-term recency.
7. O'Reilly, R. C., Braver, T. S., & Cohen, J. D. (1999). A biologically-based computational model of working memory. In A. Miyake & P. Shah (Eds.), *Models of working memory: mechanisms of active maintenance and executive control.* Cambridge: Cambridge University Press.
8. O'Reilly, R. C., & Munakata, Y. (2000). *Computational explorations in cognitive neuroscience. Understanding the mind by simulating the brain.* Cambridge, MA: MIT Press.
9. Taylor, J. G., & Taylor, N. R. (2000). Analysis of recurrent cortico-basal ganglia-thalamic loops for working memory. *Biological Cybernetics, 82*, 415-432.
10. Davelaar, E. J., & Usher, M. (2003, July). *Modeling serial position functions of neuropsychological patients.* Poster presented at the meeting of the Experimental Psychological Society. Reading, UK. Available at http://www.geocities.com/ejdavelaar
11. Murdock, B. B. (1962). The serial position effect of free recall. *Journal of Verbal Learning and Verbal Behavior, 64*, 482-488.
12. Glanzer, M. (1972). Storage mechanisms in recall. In G. H. Bower & J. T. Spence (Eds.), *The psychology of learning and motivation, Vol 5.* New York: Academic Press.
13. Philips, J. L., Shiffrin, R. M., & Atkinson, R. C. (1967). Effects of list length on short-term memory. *Journal of Verbal Learning and Verbal Behavior, 6*, 303-311.
14. Kahana, M. J. (1996). Associative retrieval processes in free recall. *Memory and Cognition, 24*, 103-109.
15. Hebb, D. O. (1949). *The organization of behavior: a neuropsychological theory.* New York: Wiley.
16. James, W. (1890). *Principles of psychology.* New York: Henry Holt.
17. Palladino, P., Cornoldi, C., De Beni, R., & Pazzaglia, F. (2001). Working memory and updating processes in reading comprehension. *Memory and Cognition, 29*, 344-354.
18. Usher, M., & Davelaar, E. J. (2002). Neuromodulation of decision and response selection. *Neural Networks, 15*, 635-645.

EFFECT OF THE LEARNING MATERIAL STRUCTURE ON RETROACTIVE AND PROACTIVE INTERFERENCE IN HUMANS: WHEN THE SELF-REFRESHING NEURAL NETWORK MECHANISM PROVIDES NEW INSIGHTS

S. C. MUSCA, S. ROUSSET AND B. ANS

Laboratoire de Psychologie et NeuroCognition — CNRS UMR 5105
Université Pierre Mendès France, BP 47, 38040 Grenoble Cedex 9, France

Following Mirman and Spivey's investigation [12], Musca, Rousset and Ans conducted a study on the influence of the nature of the to-be-learned material on retroactive interference (RI) in humans [13]. More RI was found for unstructured than for structured material, a result opposed to that of Mirman and Spivey [12]. This paper first presents two simulations. The first, using a three-layer backpropagation hetero-associator produced a pattern of RI results that mirrored qualitatively the structure effect on RI found in humans [13]. However the level of RI was high. In the second simulation the Dual Reverberant memory Self-Refreshing neural network model (DRSR) of Ans and Rousset [1, 2] was used. As expected, the global level of RI was reduced and the structure effect on RI was still present. We further investigated the functioning of DRSR in this situation. A proactive interference (PI) was observed, and also a structure effect on PI. Furthermore, the structure effect on RI and the structure effect on PI were negatively correlated. This trade-off between structure effect on RI and structure effect on PI found in simulation points to an interesting potential phenomenon to be investigated in humans.

1. Introduction

Retroactive interference (RI) is the forgetting of previously learned associations due to the learning of new ones. The RI level varies as a function of the characteristics of the learning material and of the situation the subjects are placed in [*e.g.* 3, 4, 5, 7, 8, 10, 15, 16], and supplies one with constraints to put on theoretical models of learning and forgetting. RI is generally investigated in situations involving sequential learning of two lists of items (*i.e.* associations). A first list (L1) is learned and a first test (T1) assesses subjects' memory of the learned associations. Then a second list (L2) is learned, followed by a final test (T2) on both L1 and L2 associations.

Few studies have both investigated the link between the to-be-learned material and the level of RI and tried to integrate the results into an explicit, implemented model (but see [6]). Furthermore, most of the studies on RI have been carried out with meaningful material (*i.e.* words). In this case, a strong manipulation of the structure of the learning material — opposing structured (*i.e.* rule-based) to unstructured (*i.e.* arbitrary) associations — is impossible to achieve. Indeed words convey meaning and the possibility to form associations based on meaning precludes the existence of purely unstructured associations.

Recently, Mirman and Spivey [12] investigated the effect of the nature of the learning material on RI using meaningless associations. The behavioural data pattern — from the retroactive interference paradigm they used — shows more RI for structured than for unstructured associations. To account for this result, they proposed a mixture-of-experts neural network called Dual-Strategy Competitive Learner (DSCL). DSCL works on the principle of a competition between two experts, that is "two sub-networks [...] differentially effective based on the learning task" ([12], p. 266): One, distributed, is efficient in learning rule-based items, the other, rather localized — ALCOVE [11] —, is efficient in learning arbitrary items. A crucial role is devoted to a gating network "... trained to decide which expert is the correct one for a given input; that is, which sub-network's output will be used as the overall output" ([12], p. 266).

Some shortcomings in the behavioural experiment reported in [12] cast serious doubts on the validity of the results. There are reasons to think that the items in the *Structured* condition were not learned at the exemplar level, and proactive interference was not controlled so as to equate *Structured* and *Unstructured* conditions with respect to this variable.

We elsewhere [13] discussed at length these shortcomings and provided a study conducted with the same paradigm (*cf.* Appendix for a brief description of the experimental situation and of the material used) where care was taken to eliminate them. The result was that the observed behavioural pattern [13] was *opposite* to the one reported in [12]: More RI was observed for the unstructured associations (*cf.* Figure 1a).

In the present paper we start by showing that a single system as simple as a three-layer backpropagation network can simulate this latter pattern of results. We next point at a limitation of this first simulation: The RI level is quite high when compared to the one observed in humans. The Dual Reverberant memory Self-Refreshing neural network model (DRSR) developed by Ans and Rousset [1, 2] is known to effectively reduce RI. Therefore a second set of simulations is conducted with this more complex architecture. Though made up of two parts, DRSR is to be considered as a single system since a single learning rule is always used whatever the to-be-learned material. Finally we present further results of the simulations conducted with DRSR.

2. Simulation of the Behavioural Experiment

The simulation material is the one used for the simulation presented in [13]. It corresponds to the coding of the items of the behavioural experiment

(Experiment 2 in [13]): the 17 letters used in experiment's material are coded as 17 distinct vectors, each vector containing one *one* and 16 *zeros*.

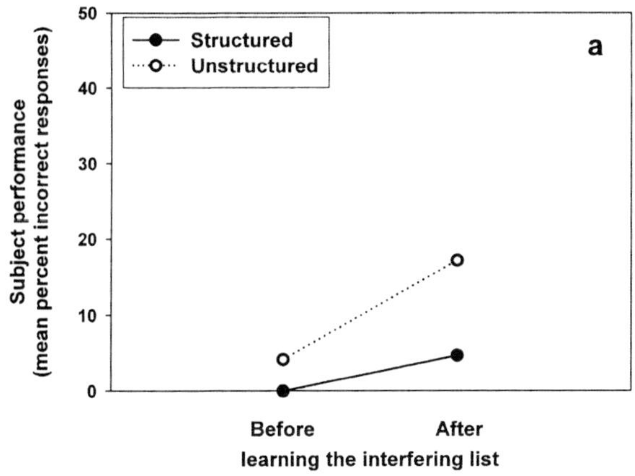

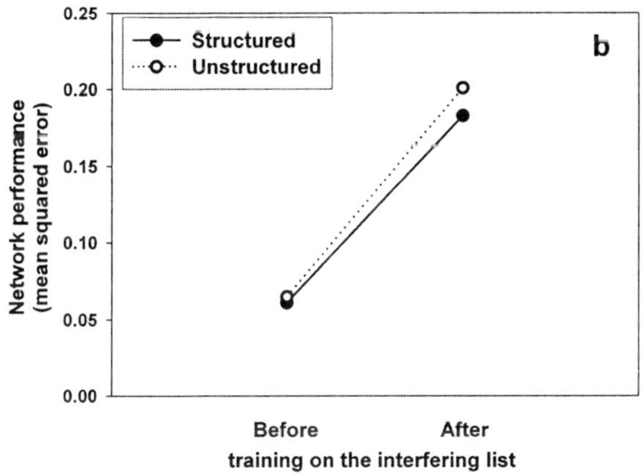

Figure 1. Performance on L1 associations before and after training on L2 associations. **a**: Behavioural experient (Experiment 2 in [13]). **b**: Simulation with the standard three-layer backpropagation hetero-associator. In both parts of the figure, the maximum on the Y-axis (50% in the behavioural experiement, and 0.25 for the simulations) corresponds to a chance-level performance, so that interference slopes can be compared. Note the steeper slope for the simulation.

2.1. *Simulations with a Standard Three-layer Backpropagation Hetero-Associator*

Does one have to assume the existence of more than one system of memory or can a single memory system account for the behavioural results of [13]? To answer this question, a simulation was run.

To allow for a comparison with simulations presented in [12], a simple hetero-associator trained with the standard backpropagation learning algorithm [14] was used. Starting with random connection weights — uniformly sampled between -0.5 and 0.5 — the network was trained with a learning rate of 0.2 and a momentum of 0.9 for 30 epochs. Twenty replications were run*. The results† qualitatively mirror those obtained in the behavioural experiment (*cf.* Figure 1), with a stronger RI for the unstructured associations [$F(1, 78) = 17.00$, $p < 0.001$]. Obviously this is opposed to the simulation result in [12]. Moreover, DSCL proposed in [12] cannot produce this result pattern, since by construction it exhibits more RI on structured associations.

However, the simulation with the hetero-associator results in quite a high level of global RI. In the following Ans and Rousset's DRSR [1, 2] will be used as it has been shown [1, 2] that it efficiently reduces the RI level in sequential learning tasks.

2.2. *Simulations with the Dual Reverberant Memory Self-Refreshing Neural Network (DRSR)*

DRSR comprises two auto-hetero-associators trained with a backpropagation learning algorithm that minimizes the cross-entropy cost function [9]. First, NET1, the "first half" of the architecture, was trained on L1 associations, then it generated pseudopatterns: Binary random input was fed to the input layer, resulting activation propagated through the network to the output layer, the output was then fed to the input layer (*re-injection* through the process of *reverberation*), and again propagated through the network, and so on. Pseudopatterns are input-output pairs with the input being taken after five re-injections. The reverberated‡ pseudopatterns (PP) generated in NET1 were used to train NET2§, the "complementary half" of the architecture.

---

* A replication consists in training 4 identical networks (*i.e.* with exactly the same random connexion weights), each one on the material of an experimental group from Experiment 2 in [13] (there were two groups per condition, *cf.* Appendix).

† In all the analyses reported from now on, PI (RI) level was controlled when comparing RI (PI) for the two experimental groups (*i.e. Structured* vs *Unstructured*), both for behavioural and for simulation results.

‡ The number of re-injections (*i.e.* reverberations) when generating a pseudopattern (*cf.* [1, 2]) was set to 5 both for NET1 and NET2.

§ NET1 was initialized with random connection weights — uniformly sampled between -0.5 and 0.5 — and trained with a learning rate of 0.01 and a momentum of 0.5 for 2000 epochs. NET2 was initialized and trained with the same parameters as NET1.

After having been trained on the PPs issued from NET1, NET2 in its turn generates PPs. These PPs are used to train NET1, which is now concurrently trained on L2 associations, so as to avoid catastrophic forgetting of the L1 associations. This is what the memory *self-refreshing* mechanism consists in. Eight replications per group were run.

In the simulations reported below, 2 parameters are manipulated: the number of pseudopatterns (*Pn*) sent from NET1 to NET2 before the latter generates PPs that are used to maintain NET1's memory of L1 (tested values: 10,000, 25,000, 50,000, 75,000 and 100,000), and the ratio (*Pr*) between the training epochs on PPs and on L2 associations in NET1 (tested values: 1, 2, and 5). For example, a *Pr* value of 2 means that any weight update (in NET1) due to the training of one L2 association was followed by weight updates due to training on two PPs. The parameters *Pn* and *Pr* correspond to the quality and the amount of the self-refreshing that allows avoiding catastrophic forgetting. In humans these parameters do not have a direct equivalent but are related to the experimental conditions: the more important the period of time between the learning of L1 and the final test, the lesser the RI.

In all the analyses reported from now on the dependent variable is the networks' performance calculated as the root mean squared error (RMS). First of all, considering all the levels of *Pn* and *Pr*, the learning material's structure effect on RI was present [$F(1, 450) = 125.22$, $p < 0.001$]: unstructured associations suffered more from RI. Thus this architecture does simulate the behavioural pattern of results of [13].

As expected, the self-refreshing efficiently reduced the RI level. Whatever the structuration level there is a significant effect of *Pn* [$F(4, 450) = 1607.57$, $p < 0.0001$] and of *Pr* [$F(2, 450) = 17.36$, $p < 0.001$]: the higher the value of each of these parameters, the lower the RI (*cf.* Figure 2). It has also been found that the effect of structure on RI depends on the parameter *Pn* [$F(4, 450) = 5.75$, $p < 0.001$]: the higher the value of *Pn*, the lower the effect of structure on RI. This was not the case for *Pr*, which we interpret as a ceiling effect.

Focusing now on proactive interference, a significant PI effect was found [$F(1, 450) = 1575.82$, $p < 0.001$]. A significant effect on the PI level was also found both for *Pn* [$F(4, 450) = 5.90$, $p < 0.001$] and for *Pr* [$F(2, 450) = 96.41$, $p < 0.001$]: the higher the value of each of these parameters, the higher the PI. The learning material's structure effect on PI was also significant [$F(1, 450) = 22.56$, $p < 0.001$]: unstructured associations suffered more from PI.

Taken together, these results show that the DRSR model allows for a fine grained simulation of the behavioural results presented in [13], but also gives rise to a structure effect on PI.

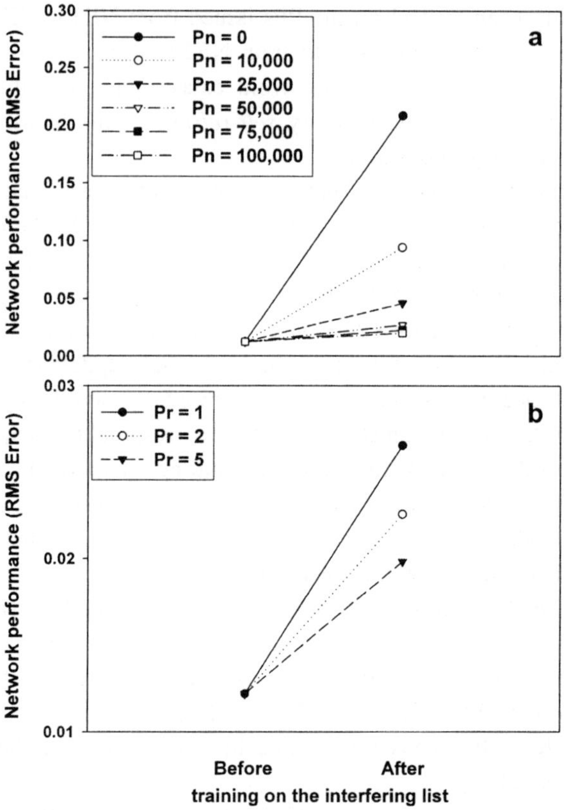

Figure 2. Performance on L1 associations before and after training on L2 associations in simulations with DRSR (*Structured* and *Unstructured* results are averaged). **a**: For $Pr = 5$, as a function of Pn. **b**: For $Pn = 100{,}000$, as a function of Pr. Note that $Pn = 0$ (in a.) corresponds to the situation where there is no memory self-refreshing, and is given here for comparison's sake. Also note a Y-axis scale change between the two parts of the figure.

So far, a structure effect has been evidenced both on RI and on PI. Is there a relationship between these effects? Are they somehow related, and what would be the direction of the correlation? When RI is reduced by self-refreshing, so is the structure effect on RI, but PI is increased. We therefore hypothesized that the structure effect on RI was negatively correlated with the structure effect on PI. This is not to say that a very high structure effect on RI (PI) could give rise to a proactive (retroactive) advantage, but merely that a high structure effect on RI (PI) could mask a — low — structure effect on PI (RI).

We therefore investigated the presence of a negative correlation between structure effect on RI and on PI (*cf.* Figure 3). Because there is no *a priori* reason that the relationship be linear, we calculated Spearman's rank correlation *rho*. Obtained values of *rho* were of -0.6 for $Pr = 1$, and of -0.9 both for $Pr = 2$ and $Pr = 5$. Thus the simulation results indicate a strong negative correlation between structure effect on RI and structure effect on PI.

3. Conclusion

An investigation of the effect of learning material's structure on RI in humans showed that subjects who learned unstructured associations lists exhibited more RI than subjects who learned structured associations lists [13].

This pattern of results was simulated with a classical three-layer hetero-associator; however the high RI level was considered as unrealistic. Simulations with Ans and Rousset's DRSR, apart from producing the expected structure effect of RI — while efficiently reducing the global RI level — showed a structure effect on PI and a very strong negative correlation between structure effect on RI and structure effect on PI. In our view this should be interpreted as the possibility for a high structure effect on RI (PI) to mask a — low — structure effect on PI (RI).

Based on this finding we hypothesize the existence of a trade-off between structure effect on RI and structure effect on PI in humans. A high structure effect on RI (PI) would mask a — low — structure effect on PI (RI). Two (complementary) analyses of the behavioural data would possibly allow for testing this hypothesis: those concerned with the two extremes of this possible trade-off. The first analysis showed a structure effect on RI [13] while the structure effect on PI was absent. It supports the trade-off hypothesis.

The second one would test whether a structure effect is found on PI in the absence of a structure effect on RI. At first glance such an analysis could be performed on the data in [13] since no structure effect on RI was found when all subjects were considered (*i.e.* when PI was not controlled but allowed to co-vary with RI). Nonetheless this analysis could be misleading because the behavioural experiment in [13] is not necessarily suited for investigating the possible structure effect on PI suggested by the simulations conducted with DRSR. Indeed that experiment was only designed to investigate in the most rigorous way the structure effect on RI.

As a conclusion, simulations conducted with DRSR not only allowed for producing a result pattern that mirrors the one observed in humans [13], but yielded a structure effect on PI and a trade-off between structure effect on RI and structure effect on PI. Unfortunately, with regard to behavioural data, the structure effect on PI was overlooked in [13] for at least two reasons. First of all, a learning material's structure effect on RI was under investigation, so structure effect on PI was considered as just a to be controlled variable. Secondly, the structure effect on PI could not have been found in the behavioural experiment in [13], because in order to control PI only the data from subjects exhibiting a nil PI was analyzed. As the experimenters' interest was in the structure effect on RI the experimental design was aimed at thoroughly testing for such an effect; it was not suited for testing for the existence of a structure effect on PI.

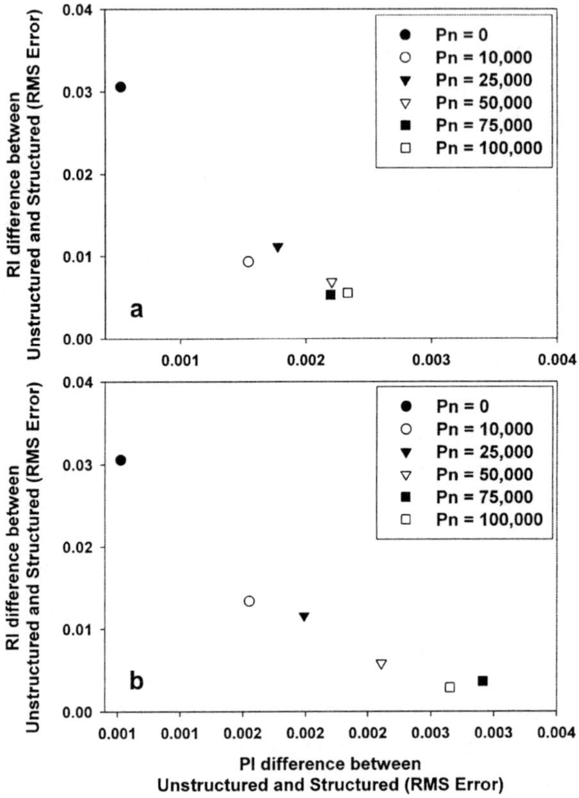

Figure 3. Structure effect on RI as a function of structure effect on PI (in simulations with DRSR). **a**: For $Pr = 1$. **b**: For $Pr = 5$.

Further behavioural experiments are needed to investigate the existence of a learning material's structure effect on PI, and of a trade-off between structure effect on RI and structure effect on PI as suggested by the simulations conducted with Ans and Rousset's DRSR.

Acknowledgements

This work was supported in part by a research grant from the European Commission (HPRN-CT-1999-00065) and by the French government (CNRS UMR 5105).

Appendix

	Structured					*Unstructured*		
	1st part	1st choice	Rule	2nd choice	Rule	1st part	1st choice	2nd choice
List 1 (L1)	RIFO	ROFI	R1	IFOR	R3	RIFO	VIBA	ANED
	VAPI	VIPA	R1	APIV	R3	VAPI	ECAN	ELAB
	TOVE	TEVO	R1	OVET	R3	TOVE	REFO	IZAD
	VACE	VECA	R1	VOCI	R4	VACE	EGAT	PAZI
	PAGI	PIGA	R1	POGE	R4	PAGI	EZAD	GAFI
	FOVI	FIVO	R1	FAVE	R4	FOVI	TIBA	ZACE
	LAFE	ELAF	R2	AFEL	R3	LAFE	RITO	AZEB
	DATI	IDAT	R2	ATID	R3	DATI	IPAM	REDA
	MOZE	EMOZ	R2	OZEM	R3	MOZE	FOZE	AVER
	CATE	ECAT	R2	COTI	R4	CATE	DIZO	AFEP
	CAGI	ICAG	R2	COGE	R4	CAGI	ACEP	NEFA
	POBE	EPOB	R2	PABI	R4	POBE	GIDO	OGEP
List 2 (L2)	BAGE	AGEB	R3	BEGA	R1	BAGE	OLER	LACI
	ZELA	ELAZ	R3	ZALE	R1	ZELA	BOVI	DOFE
	ZEVO	EVOZ	R3	ZOVE	R1	ZEVO	OCIB	ORIL
	FIGA	FEGO	R4	FAGI	R1	FIGA	VEDA	ABER
	VIZO	VEZA	R4	VOZI	R1	VIZO	DACI	AVEF
	GOPE	GAPI	R4	GEPO	R1	GOPE	CEDO	NATI
	NEDO	EDON	R3	ONED	R2	NEDO	IDAG	AFEG
	CINA	INAC	R3	ACIN	R2	CINA	VAFI	IZOM
	NIZO	NEZA	R3	ONIZ	R2	NIZO	ENOP	OTIG
	LENO	LINA	R4	OLEN	R2	LENO	OTEB	TARI
	RILA	RELO	R4	ARIL	R2	RILA	PIBO	CAPI
	MIPO	MEPA	R4	OMIP	R2	MIPO	ILOB	FENO

Example of the experimental material used in [13] for a *Structured* and an *Unstructured* groups (there were two more groups, for list counterbalancing sake). Each item is made of two parts ("1st part" and "1st choice" in the table below ; *e.g.* RIFO ROFI, VAPI VIPA, *etc.*). Subjects were to learn the items of L1, were tested in a two-choice recognition task (with the corresponding

"2nd choice" as a distractor), then learned L2 items (*e.g.* BAGE AGEB, ZELA ELAZ, *etc.*), and were finally tested in a two-choice recognition task on both L1 and L2 items.

The rules for constructing each of the *Structured* condition item are mentioned (R1 to R4, one can easily deduce from the given examples what these rules are). As for the *Unstructured* condition, the items' parts were paired at random to create the experimental material.

References

1. Ans, B., & Rousset, S. (1997). Avoiding catastrophic forgetting by coupling two reverberating neural networks. *Comptes Rendus de l'Académie des Sciences Paris, Sciences de la vie, 320*, 989-997.
2. Ans, B., & Rousset, S. (2000). Neural networks with a self-refreshing memory: Knowledge transfer in sequential learning tasks without catastrophic forgetting. *Connection Science: Journal of Neural Computing, Artificial Intelligence and Cognitive Research, 12*(1), 1-19.
3. Barnes, J. M., & Underwood, B. J. (1959). "Fate" of first-list associations in transfer theory. *Journal of Experimental Psychology, 58*, 97-105.
4. Bäuml, K.-H. (1996). Revisiting an old issue: Retroactive interference as a function of the degree of original and interpolated learning. *Psychonomic Bulletin & Review, 3*, 380-384.
5. Bunt, A. A., & Sanders, A. F. (1972). Some effects of cognitive similarity on proactive and retroactive interference in short-term memory. *Acta Psychologica, 36*(3), 190-196.
6. Chappell, M., & Humphreys, M. S. (1994). An auto-associative neural network for sparse representations: Analysis and application to models of recognition and cued recall. *Psychological Review, 101*(1), 103-128.
7. Cofer, C. N., Faile, N. F., & Horton, D. L. (1971). Retroactive inhibition following reinstatement or maintenance of first-list responses by means of free recall. *Journal of Experimental Psychology, 90*(2), 197-205.
8. Delprato, D. J. (1970). Successive recall of List 1 following List 2 learning with two retroactive inhibition transfer paradigms. *Journal of Experimental Psychology, 84*(3), 537-539.
9. Hinton, G. E. (1989). Connectionist learning procedures. *Artificial Intelligence, 40*, 185-234.
10. Izawa, C. (1980). Proactive versus retroactive interference in recognition memory. *Journal of General Psychology, 102*(1), 53-73.
11. Kruschke, J. K. (1992). ALCOVE: An exemplar-based connectionist model of category learning. *Psychological Review, 99*(1), 22-44.
12. Mirman, D., & Spivey, M. (2001). Retroactive interference in neuronal networks and in humans: The effect of pattern-based learning. *Connection-Science, 13*(3), 257-275.

13. Musca, S. C., Rousset, S., & Ans, B. (submitted). Differential forgetting of structured and unstructured associations in artificial neural networks and humans: Structured associations resist better.
14. Rumelhart, D. E., Hinton, G. E., & Williams, R. J. (1986). Learning Internal Representations by Error Propagation. In D. E. Rumelhart & J. L. McClelland (Eds.), *Parallel Distributed Processing - Explorations in the Microstructure of Cognition* (Vol. 1, pp. 318-362). Cambridge, MA: MIT Press.
15. Shulman, H. G., & Martin, E. (1970). Effects of response-set similarity on unlearning and spontaneous recovery. *Journal of Experimental Psychology, 86*(2), 230-235.
16. Wheeler, M. A. (1995). Improvement in recall over time without repeated testing: Spontaneous recovery revisited. *Journal of Experimental Psychology: Learning, Memory, and Cognition, 21*(1), 173-184.

Vision

SPATIOTEMPORAL LINEAR SIMPLE-CELL MODELS BASED ON TEMPORAL COHERENCE AND INDEPENDENT COMPONENT ANALYSIS

JARMO HURRI[A], JAAKKO VÄYRYNEN[A] AND AAPO HYVÄRINEN[B,A]

[A] *Neural Networks Research Centre*
Helsinki University of Technology, P.O.Box 9800, 02015 HUT, Finland

[B] *Helsinki Institute for Information Technology / BRU*
Department of Computer Science
University of Helsinki, P.O.Box 26, 00014 UH, Finland

The search for computational principles that underlie the functionality of different cortical areas is a fundamental scientific challenge. In the case of sensory areas, one approach is to examine how the statistical properties of natural stimuli – in the case of vision, natural images and image sequences – are related to the response properties of neurons. For simple cells, located in V1, the most prominent computational theories linking neural properties and stimulus statistics are temporal coherence and independent component analysis. For these theories, the case of spatial linear cell models has been studied in a number of recent publications, but the case of spatiotemporal models has received fairly little attention. Here we examine the spatiotemporal case by applying the theories to natural image sequence data, and by analyzing the obtained results quantitatively. We compare the properties of the spatiotemporal linear cell models learned with the methods against each other, and against parameters measured from real visual systems.

1. Introduction

Simple cells, located in the primary visual cortex, are selective for a number of visual stimulus characteristics. Many of these neurons respond strongest when the visual stimulus is at a specific location in the visual field, has a certain orientation and frequency, and moves in a particular direction. Why is this the case?

If one builds an information processing system that is optimal for certain type of input, the properties of the input will be reflected in the resulting system. The visual system of an adult animal has been shaped by forces of evolution, and self-organization during development. Saying that the visual system is optimal for processing typical visual stimuli would probably be too strong a statement, because evolution is a greedy process that builds incrementally upon earlier solutions, but perhaps it can be considered to be close to optimal. This type of reasoning has led to the idea that the statistical properties of natural visual stimuli have influenced the functional properties of cells in the primary visual cortex (for a review, see Ref. 1). The search for "optimality principles" in the visual cortex has resulted in some influential computational theories, including temporal coherence[2,3,4], and the (closely related) theories of sparse coding and independent component analysis (ICA).[5,6,7]

Studies in this field typically employ linear filters as models of the classical receptive fields (CRFs) of simple cells. The classical spatial receptive field is a description of how the neuron responds to localized onsets and offsets of light inside that portion of the visual field where the neuron can be excited. The classical *spatiotemporal* receptive field includes also a description of the dynamics of these responses. The case of spatial receptive-field models has been studied extensively for both temporal coherence and ICA/sparse coding (e.g., Refs. 5, 4). Some studies employing spatiotemporal models in ICA and sparse coding have also been published.[8,9] However, in the case of temporal coherence, no studies employing spatiotemporal CRFs have been reported. Consequently, no comparisons between the spatiotemporal results obtained with temporal coherence and ICA/sparse coding have been possible. Also, none of the previously published reports on spatiotemporal models consider the most comprehensive physiological measurements of spatiotemporal simple-cell receptive fields.[10]

In this paper, we use temporal coherence and independent component analysis to learn spatiotemporal CRFs from natural image sequence data. There are three main contributions in this paper: the qualitative and quantitative description of temporally coherent spatiotemporal CRFs, the comparison of results obtained with temporal coherence and independent component analysis against each other, and the comparison of the results obtained with these methods against recent, comprehensive physiological data. In what follows, we first describe the measure of temporal coherence used in this paper in Section 2. Independent component analysis is a well-known method (see, e.g., Ref. 7), so it will not be explained here. In Section 3 we describe the data and preprocessing used in the experiments. The obtained results are described and analyzed quantitatively in Section 4, where we also provide a discussion of the differences between our results and neurophysiological measurements. Finally, we conclude this paper with conclusions in Section 5.

2. Temporal coherence of activity levels

The core idea of temporal coherence is that the neural representation changes as little as possible over time, while still preserving (almost) all of the information about the input data. It has previously been shown that maximization of *temporal coherence of activity levels* is one computational principle which leads to the emergence of simple-cell-like spatial CRFs from natural image sequences.[4] The exact mathematical definition of this principle is given below.

In this paper we study the case of spatiotemporal CRFs, presented here in a matrix-vector formulation. A *vectorization* of spatiotemporal image sequence samples can be done by scanning the frames of an image sequence one by one column-wise into a vector. Let a vectorized sequence of 8 frames of size 11×11 pixels, taken from natural video at time t, be denoted by the 968-dimensional ($= 8 \times 11^2$) vector $\mathbf{x}(t)$. Let $\mathbf{y}(t) = [y_1(t) \cdots y_K(t)]^T$ represent the outputs of K simple cells. In the standard linear filtering model, $\mathbf{y}(t) = \mathbf{W}\mathbf{x}(t)$. In this model, the set of filters (vectors) $\mathbf{w}_1, ..., \mathbf{w}_K$ corresponds to the spatiotemporal receptive fields of simple cells, and $\mathbf{W} = [\mathbf{w}_1 \cdots \mathbf{w}_K]^T$ denotes a matrix with all the filters as rows.

Temporal response strength correlation[4], the objective function, is defined by

$$f(\mathbf{W}) = \sum_{k=1}^{K} \mathrm{E}_t \left\{ g(y_k(t)) g(y_k(t - \Delta t)) \right\}, \tag{1}$$

where the nonlinearity g is strictly convex, even (rectifying), and differentiable, and Δt denotes a delay in time. Examples of choices for the nonlinearity are $g_1(x) = x^2$ and $g_2(x) = \ln \cosh x$. A set of CRFs which has a large temporal response strength correlation is such that the same neurons *often respond strongly at consecutive time points*, outputting large (either positive or negative) values, thereby expressing temporal coherence of the activity of populations of neurons. Additional constraints are used to keep the outputs of the neurons bounded and to keep the CRFs from converging to the same solution, and a gradient projection method can be used to maximize the resulting constrained optimization problem.[4] The initial value of $\mathbf{W}$ is selected randomly.

One standard way to interpret the results obtained with linear simple-cell models is to express the relationship between data $\mathbf{x}(t)$ and neural responses $\mathbf{y}(t)$ as a *generative model*[5,11,9]: $\mathbf{x}(t) = \mathbf{A}\mathbf{y}(t)$. If $\mathbf{x}(t)$ and $\mathbf{y}(t)$ have the same dimension, then $\mathbf{A} = \mathbf{W}^{-1}$, otherwise $\mathbf{A}$ can be obtained by computing the pseudoinverse solution. In the generative-model formulation, each column of matrix $\mathbf{A}$ can be interpreted as the feature coded by the corresponding simple cell. Below we will use this interpretation when we present our results.

3. Data collection and preprocessing

The data used in the experiments were sampled from the database of natural image sequences described in Ref. 8. The sampled data consisted of 120,000 image sequence blocks of size $11 \times 11 \times 9$, where the first two dimensions denote spatial size and the last dimension denotes length in time. Each sample of length 9 time frames was divided into two partially overlapping samples of length 8 frames; this yields $\mathbf{x}(t)$ and $\mathbf{x}(t-\Delta t)$. The sampling frequency of the data was 25 Hz, so Δt was 40 ms, and the durations of $\mathbf{x}(t)$ and $\mathbf{x}(t-\Delta t)$, and the spatiotemporal CRF, were 280 ms. Preprocessing consisted of removal of local average image intensity (DC component of the spatiotemporal block) and dimensionality reduction by 50% to 484 using principal component analysis.[7] Dimensionality reduction reduces the effect of noise and aliasing artifacts, and decreases the computational complexity of the problem (the degree of dimensionality reduction applied here retains 95% of original signal energy).

4. Results and discussion

Some of the resulting spatiotemporal basis vectors of size $11 \times 11 \times 8$ (i.e., columns of $\mathbf{A}$) maximizing the objective function (Eq. 1) are shown in Fig. 1. As can be seen, the learned receptive fields share the primary spatial properties of simple cells in that they are localized, oriented, and have multiple scales (see, e.g., Ref. 12). In addition to these spatial properties, the receptive fields also have physiologically relevant qualitative temporal properties. Some of the receptive fields seem to be space-time separable[10], while others are inseparable.[a] Some of the separable receptive fields have constant time profiles, some of them have changing time profiles. Also, different space-time inseparable receptive fields seem to respond to different velocities.

To obtain a corresponding set of ICA results, we applied the symmetric fast fixed-point ICA algorithm[7] with nonlinearity $g(y) = \tanh y$

[a] A space-time separable receptive field can be expressed as a product of a one-dimensional temporal profile and a two-dimensional spatial profile.

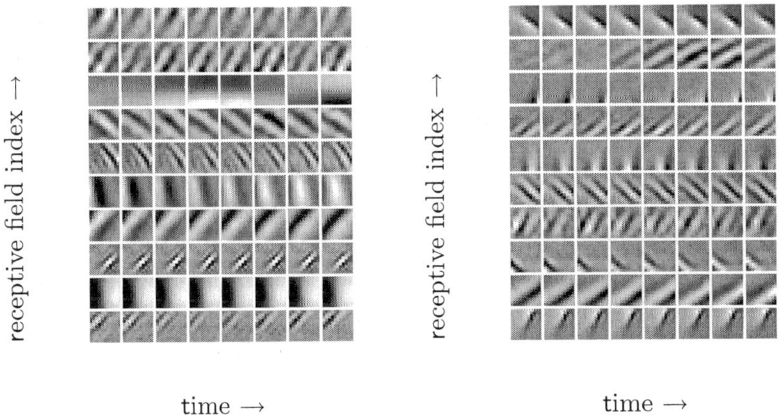

Figure 1. A subset of 20 spatiotemporal receptive field models (columns of **A**) obtained by maximizing temporal coherence of activity levels in natural image sequences (10 receptive fields in the image on the left and 10 on the right). Each of the 20 rows corresponds to one spatiotemporal receptive field model, and the frames in a row correspond to spatial snapshots of a spatiotemporal receptive field at consecutive time instances.

to the same data.

In order to assess the results quantitatively, we measured some important parameters from the two sets of CRFs obtained using temporal coherence and ICA. All of the steps in the following are as in Ref. 10. The three-dimensional x–y–t descriptions of the receptive fields were first reduced to two-dimensional x–t profiles by rotating them so that the y-dimension coincided with the preferred orientation of the receptive field, and then integrating (summing) the three-dimensional data along the y-axis. The amplitude spectrum of this x–t profile was then taken to provide a frequency domain description of the receptive field. Three parametric curves were fit to the resulting spatial and frequency descriptions. A one-dimensional Gabor function was fit to the x–t profile at the t-coordinate where the one-dimensional (spatial) slice had maximum overall area (this time point t is defined as peak response latency). A product of a Gaussian and a gamma distribution was fit to the positive quadrant of the two-dimensional amplitude spectrum; another function of the same form was fit to a second quadrant (only two quadrants need to be considered because of symmetry of the amplitude spectrum). Below we will refer to the combination of the last two curves as the amplitude spectrum curve. A temporal profile of the receptive field was obtained by taking a temporal slice through the maximum absolute value of the receptive field. In addition, the *envelope* of the temporal profile was determined as a basis for description of the duration of the receptive field.

The measured parameters were

- optimal spatial frequency (spatial frequency coordinate of the peak of the amplitude spectrum curve; Figure 2A)
- high spatial frequency cutoff (high-frequency point above optimal spatial frequency where amplitude spectrum curve has dropped to half of the maximum; Figure 2B)
- envelope width (width parameter of the Gabor function; Figure 2C)
- number of subregions (number of separate spatial dark and bright regions, computed from the optimal spatial frequency and envelope width; Figure 2D)
- spatial phase (phase parameter of the Gabor function; Figure 2E)
- receptive field duration (the width of the envelope of the temporal profile above $1/e$ of the peak envelope value; Figure 2F)
- peak response latency (see text above; Figure 2G)
- optimal temporal frequency (defined similarly as optimal spatial frequency; Figure 2H)
- high temporal frequency cutoff (defined similarly as high spatial frequency cutoff; Figure 2I)
- direction selectivity index (a measure of whether the filter responds best to stationary or moving stimuli; determined from the symmetry/asymmetry of the amplitude spectrum; Figure 2J).

The exact operational definitions of these quantities can be found in Ref. 10. Note that spatial measurement information is shown in Figs. 2A–E, while Figs. 2F–J contain temporal measurement information. As a general observation, the histograms of measured parameters for temporal coherence and ICA are mostly similar. Because of this similarity, in what follows we focus on the comparison against physiological measurements. These measurements were made by DeAngelis et al.[10] from 91 simple cells. The histograms of the measurements are shown in Fig. 3. Because of the relatively low number of measurement points (91 cells), we will only consider the distributions qualitatively, which still turns out to produce some interesting results.

When the results obtained with temporal coherence and ICA are compared against the physiological measurements, we see similarities in the spatial measurements, and major differences in the temporal measurements.

In the spatial measurements (Figs. 2A–E and 3A–E), the distributions have similar qualitative properties, although the number of subregions is substantially higher in the computational results. In the case of temporal measurements (Figs. 2F–J and 3F–J), the physiological measurements are strikingly different from the measurements of learned CRFs. In all other cases except direction selectivity, the histograms of the physiological measurements look almost completely different from the corresponding histograms obtained from the two models. One key measurement in understanding these differences is receptive field duration (Figs. 2F and 3F). A clear majority of the CRFs which emerge from the two models span the whole time frame of the receptive field (see also Fig. 1). When a CRF has a practically constant magnitude over the whole time frame, the point where the maximum is reached is somewhat arbitrary (Fig. 2G). The differences in optimal temporal frequency (Figs. 2H and 3H) are probably also related to the lack of temporal change.

Can we find any reason underlying the long duration of the learned CRFs? In this paper, we have applied so called *symmetric* algorithms in the case of temporal coherence and ICA. In these algorithms, the optimization of the objective function is done simultaneously for all the CRFs (the whole matrix $\mathbf{W}$). This method can be contrasted with *deflationary* methods, in which CRFs (rows of $\mathbf{W}$) are learned one by one, and the first solutions dominate. In general, the symmetric algorithms should be able to find a better balanced set of basis vectors.[7] The ICA results become more temporally localized if a *deflationary* algorithm – in which the filters are extracted one by one[7] – is used and dimensionality reduction is applied to the data. This observation is in concordance with the results obtained by van Hateren and Ruderman.[8] An analogous change in the algorithm and preprocessing methods improves slightly the temporal localization of results obtained by maximizing temporal response strength correlation, but not to the same degree as in the case of ICA. So far we have been unable to pinpoint the reason for this difference, and further research is needed to clarify the issue.

5. Conclusions

In this paper, we have applied temporal coherence and independent component analysis to learn spatiotemporal receptive field models from natural image sequences. Our results show that the results obtained with these two methods have similar spatial and temporal quantitative properties.

When compared with physiological measurements from cat simple cells, similarities between the learned CRFs and physiological measurements are observed in the spatial domain, while there are substantial differences in the temporal domain. Except for measurements of direction selectivity, the physiologically measured temporal parameter distributions – such as duration, peak response latency and optimal temporal frequency – are qualitatively different from the parameter distributions of learned CRFs. The reasons underlying this discrepancy are currently unknown, and pose an important problem for future research.

References

1. Eero P. Simoncelli and Bruno A. Olshausen. Natural image statistics and neural representation. *Annual Review of Neuroscience*, 24:1193–1216, 2001.
2. Peter Földiák. Learning invariance from transformation sequences. *Neural Computation*, 3(2):194–200, 1991.
3. Laurenz Wiskott and Terrence J. Sejnowski. Slow feature analysis: Unsupervised learning of invariances. *Neural Computation*, 14(4):715–770, 2002.
4. Jarmo Hurri and Aapo Hyvärinen. Simple-cell-like receptive fields maximize temporal coherence in natural video. *Neural Computation*, 15(3):663–691, 2003.
5. Bruno A. Olshausen and David Field. Emergence of simple-cell receptive field properties by learning a sparse code for natural images. *Nature*, 381(6583):607–609, 1996.
6. Anthony Bell and Terrence J. Sejnowski. The independent components of natural scenes are edge filters. *Vision Research*, 37(23):3327–3338, 1997.
7. Aapo Hyvärinen, Juha Karhunen, and Erkki Oja. *Independent Component Analysis*. John Wiley & Sons, 2001.
8. J. Hans van Hateren and Dan L. Ruderman. Independent component analysis of natural image sequences yields spatio-temporal filters similar to simple cells in primary visual cortex. *Proceedings of the Royal Society of London B*, 265(1412):2315–2320, 1998.
9. Bruno A. Olshausen. Sparse coding of time-varying natural images. In Petteri Pajunen and Juha Karhunen, editors, *Proceedings of the Second International Workshop on Independent Component Analysis and Blind Signal Separation*, pages 603–608, 2000.
10. Gregory C. DeAngelis, Izumi Ohzawa, and Ralph D. Freeman. Spatiotemporal organization of simple-cell receptive fields in the cat's striate cortex. I. General characteristics and postnatal development. *Journal of Neurophysiology*, 69(4):1091–1117, 1993.
11. Geoffrey E. Hinton and Zoubin Ghahramani. Generative models for discovering sparse distributed representations. *Philosophical Transactions of the Royal Society B*, 352(1358):1177–1190, 1997.
12. Stephen E. Palmer. *Vision Science – Photons to Phenomenology*. The MIT Press, 1999.

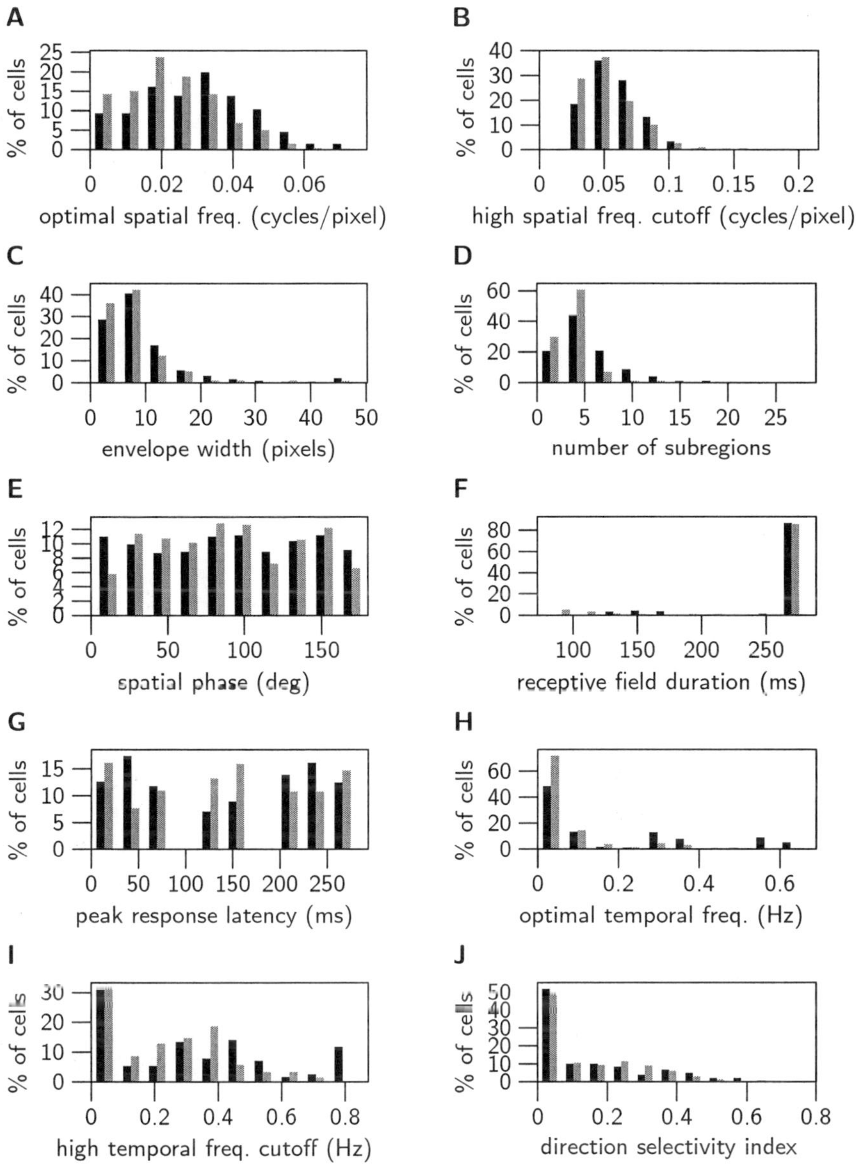

Figure 2. Quantitative measurements of spatiotemporal classical receptive fields obtained with temporal coherence (black bars) and ICA (grey bars).

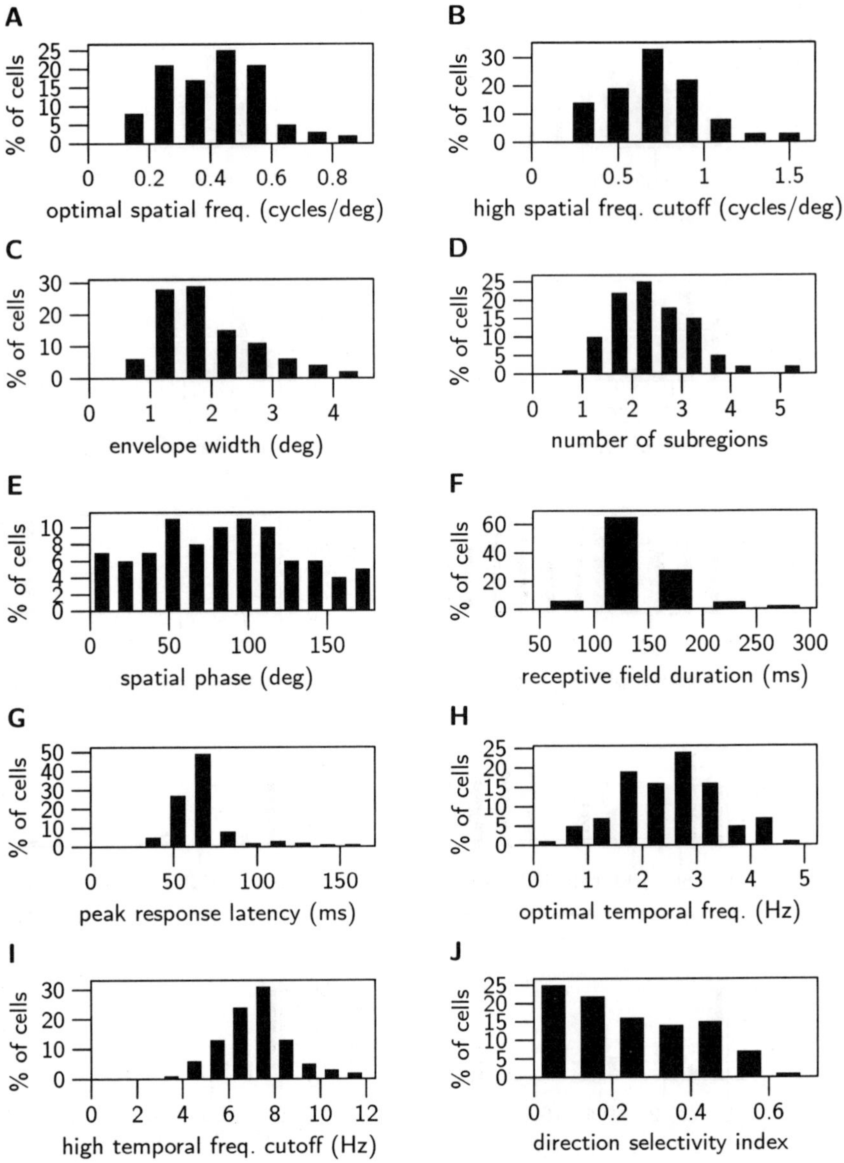

Figure 3. For comparison, results reported by DeAngelis et al.: physiological spatiotemporal receptive field measurements made from adult cats.[10] Note that in subfigures A–C, x-axis units differ from Figs. 2A–C, and that in subfigures D and F–I, x-axis limits differ from Figs. 2D and 2F–I.

PREDICTING COLLISION: A CONNECTIONIST MODEL

J. KARANKA AND D. LUQUE
University of Malaga.
University Campus of Teatinos,
Málaga, 29071, Spain
E-mail: jonikaranka@mixmail.com

There have been many proposals of how time-to-collision is computed (see Sun & Frost [1] for a review). But the results of different tasks were not conclusive for any of these models. According to new evidence of development and tuning of tasks, we propose a simple recurrent neural network [2] to account for these phenomena. Specifically we simulated ontogenic development and tuning to speed ranges through training. Results were similar to human performance: less-trained-networks responses consistently anticipate to slow objects or large objects, and this behaviour diminishes with training.

1. Introduction

1.1. *Predicting collision*

James Tresilian defined time-to-collision (TTC) as "...the time remaining before some spatial coincidence event such as the collision of the two objects or a moving objects arrival at a specific position in space" [3]. Although this definition is very general, we focus on the case of head-on approach. Objects on head-on collision trajectory increase their visual angle evenly in width and height. And many species are sensitive to the image of objects looming toward them. Some of them include fiddler crabs, fishes, frogs, turtles, chicks, monkeys and humans [1]. Possessing this skill is useful for a large number of actions e.g. avoiding collisions, precise braking or landing and catching fast moving objects. In a demanding environment, these abilities are necessary for survival in most animal species.

1.2. *Two methods to compute TTC*

Originally two types of methods have been proposed to compute TTC [4]. The first one was the cognitive method, in which TTC information is derived from information computed previously like speed and distance. The second one is the optic flow method, where TTC is computed without employing previously computed information, but directly from the changing optic array. Evidence points to the last method, showing that humans can directly compute TTC from optic flow [4, 5]. So, no contextual cues are necessary to predict with enough precision TTC. This optic flow approach has been reinforced by the recent discovery of neurons responsive to TTC in pigeons [1].

1.3. *Time-to-collision computed through optical parameters*

There are two 'optical primitives' found in the optic flow; both of which can be combined to compute TTC. The first primitive is the visual angle (θ) of the object on the retina. As the object approaches the animal, this angle increments non-linearly and the form of the curve depends on the speed and size of the object. The same is true for the second optical primitive: rate of expansion[1] (θ'). Rate of expansion is the difference between the visual angle in the current and the preceding moments. When the rate of expansion is plotted, it also increases in a non-linear way, but it increments more steeply than visual angle at the moments preceding collision. The combination of the preceding primitives result in three parameters useful to compute TTC: rho, tau and eta.

Rho is triggered when the rate of expansion reaches a critical value. Rate of expansion alone can be used to compute TTC responses because, contrary to visual angle, it yields a value of 0 for non-moving objects, and is not so sensitive to object size and speed. Some studies present data favourable to this rho-hypothesis like errors associated with speed of closure [4, 6] and size of the approaching object [7, 8].

Tau is an invariant to predict TTC. This means that if humans are capable of extracting this parameter from optic flow (where both of its components are available) their TTC predictions would not be flawed by the speed or size of the objects. Tau is usually expressed as:

$$\tau = \theta / \theta' \qquad (1)$$

There has been a long experimental tradition favouring tau [9-11], but actual interpretation of the data [12] invalidates some of these claims. Even so, psychophysical data by Regan & Hamstra [11] show that humans are capable of employing tau independently of other cues (e.g. rate of expansion).

Eta [13] has been proposed to describe the response of locust descending contralateral movement detector (DCMD) neurons to approaching, receding and translating objects. It employs two constant parameters to control the magnitude of the response (C) and the inhibitory damping of the peak (α). Eta is expressed as:

$$\eta = C \times \theta' \times e^{-\alpha\theta} \qquad (2)$$

---

[1] Rate of expansion is computed from visual angle. Although the former is not strictly a primitive we will consider it as such because it is used for computing all other optical parameters.

Eta actually provides a biologically plausible description of how a neuron peaks when anticipating collision with an approaching object. Eta is modulated by the speed and size of the approaching object, and hence, its predictions are similar to those of rho.

1.4. *The problem of adaptation*

Actually there is agreement that different sources of information are used to solve TTC tasks [1, 3]. This is concluded from the fact that no parameter can explain all the experimental data. Depending on the characteristics of the task and the amount of training, subjects seem to adapt, giving more tau-like responses in the most demanding environments, and more rho-like responses in less demanding environments [14]. Actually Tresilian identified a number of requirements for the task in order to test the tau-hypothesis [3]. All of these requirements stressed the difficulties of the task, e.g.: short time interval, fast objects, specific response methods. When these requirements were met, the results have been favorable to the tau-hypothesis (at least, in those studies in which the amount of training have not been manipulated), whereas if they were not met, the results have been more favorable to the rho-hypothesis [3]. The differences found between subjects with different amounts of training show that only the most trained subjects respond according to the predictions of the tau hypothesis [14, 15]. Neither model can explain how the transition from one strategy to the other one is performed.

Connectionist modeling seems like a plausible method to create an adaptive model to explain these facts. A neural network can extract information from optic flow employing collision feedback as teaching input, and develop through different 'stages' as it tests the different 'local minima' of the error surface. We think that this framework may be able to explain differences due to overall tuning (amount of training) and adaptation to specific tasks. In fact, a recent article by Smith, Flach, Dittman & Stanard [14] refers to the performance of subjects exposed to a task with fast speeds as adapted to a 'local minima', in comparison with a group of subjects exposed to a task with slow speeds. We can see no other way of explaining this between groups difference but by proposing an adaptive learning mechanism to account for TTC.

In this connectionist framework, we describe a neural network capable of learning a TTC task, that shows rho-like performance when it is less-trained and develops a more tau-like performance when it is trained further, and compare its performance to behavioural experiments in humans.

1.5. *Architecture of the model*

We trained a simple recurrent neural network (SRNN) employing simulated approaching objects. The SRNN is composed of three layers of units: input, hidden and output, with the hidden units connected to a special layer of 'context units'. There is a context unit for every hidden unit, and they store the hidden unit's last activation. Once the network is 'running', the hidden layer can extract information 'through time' by integrating the activation presented in the context units (for more about SRNN see Elman [2]). It is usual to train the SRNN on prediction tasks in which the network has to predict the next moment of a sequence, with its teaching input being the input of the next time step. In our case, the SRNN did not have to predict sequences, instead, it only had one output unit which coded 'collision' (see fig. 1), and had a teaching input that fedback collision information (collision or absence of collision). This was done to create an ecologically plausible[2] and simplest network for the task.

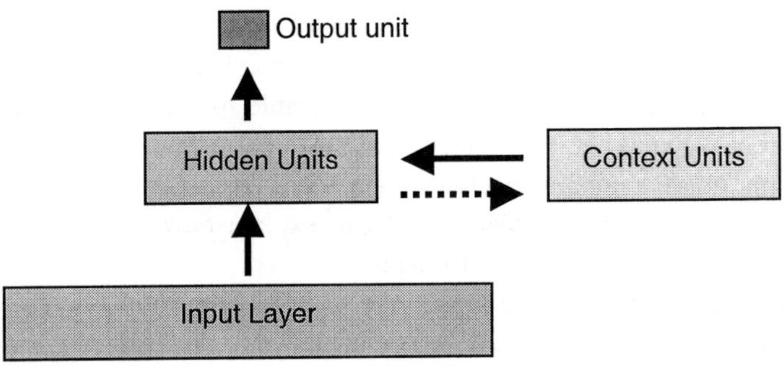

Figure 1. Architecture of the SRNN. The input layer contains 40 nodes, both hidden and context layers contain 20 nodes, and there is only one output unit. Dotted line arrows mean "The weights are fixed, the connections copy the units activity" while solid line arrows mean "Connections subject to weight change".

The objects are represented in successive temporal steps in a one-dimensional retina of 40 nodes. The visual angle in the retina depends always on distance and real size of the object and can be measured by:

$$\text{Sin}\alpha = \text{Size}/\sqrt{(\text{Size}^2 + \text{Distance}^2)} \qquad (3)$$

[2] No processes were added (e.g. computation of distance or speed) that were not presented in the non-simulated task.

The SRNN receives, for every object, visual angles until the distance reaches 0, moment at which it also receives a collision input (teaching input).

Objects were created employing two parameters: real size and speed, combined orthogonally. 12 different speeds and 16 different sizes amounted up to of 192 different looming stimuli. For all of them, initial distance is constant.

2. Simulation 1: Speed and Size Effects

2.1. *Task*

The network was trained with the 192 different stimuli (12 speeds x 16 sizes) for 75, 150 and 225 epochs. Five simulations were performed.

Once trained, the network was tested for speed and size effects. When size was kept constant speed was modified, and vice versa. Amount of training was an independent variable. Generalization tests were made by creating new objects in the size and speed range of the preceding objects, keeping size or speed constant.

2.2. *Results*

Responses were classified as anticipations, correct responses or late responses, depending on when the output unit's activation crossed a threshold[3]. When completely trained, the network had an overall of 80.72% of correct responses. We employed Kruskal-Wallis for all the subsequent analyses.

2.2.1. *Size effects*

There were more anticipation responses for large than small objects (see fig. 2), being this effect significant ($p < 0.001$). This size-dependent response was modulated by amount of training, the anticipation responses being less frequent in the most trained group ($p < 0.001$).

2.2.2. *Speed effects*

The speed effects were also significant ($p < 0.001$), with more anticipation responses for the slower objects than for the faster ones (see fig. 2). Contrary to our expectations -and empirical results- training effects were not found.

2.2.3. *Generalization test*

---

[3] We considered a response as a collision response when output unit activation surpassed 0.100. Object approaches without collision responses were considered late responses.

For the generalization task the most trained networks showed a 61.3% of correct responses. In the task, speed and size effects were replicated ($p < 0.001$, for both). This time, training effects were significant in sizes ($p = 0.006$) and speeds ($p = 0.001$).

2.3. *Discussion*

General effects found in humans were replicated, even in generalization tasks. The pattern of data reflects more anticipation for slower speeds [4, 6] and larger sizes [7, 8]. The training produced the expected dynamic behaviour reported in recent research [14, 15] and that has been explained by the use of different strategies (see general discussion). The training improves the performance of the network in the generalization tasks. This supports the idea that it is changing the strategy and not only learning to respond to concrete stimuli.

The training effects in the speed test were not found, contrary to human behavioural results [14]. This can be interpreted as a failure of the model to simulate human cognitive processes or as a methodological error, e.g. an inadequate selection of stimuli or an incorrect threshold. Regarding the remaining results, we are mostly persuaded by the methodological hypothesis.

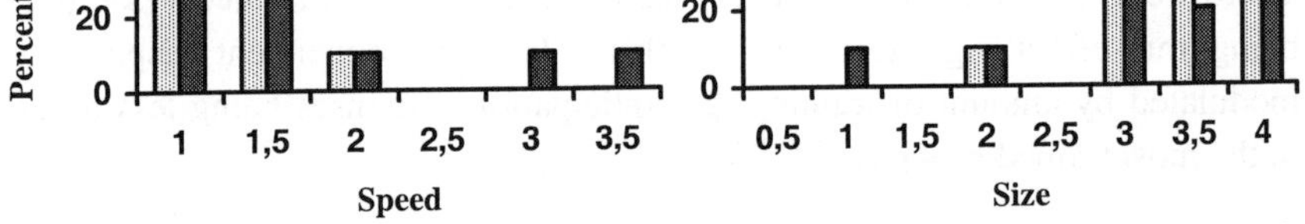

Figure 2. Results of simulation 1. Dotted bars represent responses of less-trained-networks and dark bars represent responses for more-trained-networks. The left figure represents percent of anticipation responses for different speeds, and the right figure represents percent of anticipation responses for different sizes.

3. Simulation 2: Adaptation to Speed Ranges

3.1. *Task*

In simulation 2 we replicated experiment 3 of Smith et al. [14] They argued that the subjects can modify their response strategy in "optical state space"[4] to adapt their answers to the demands of the task. Thus, if two groups of subjects are exposed to different ranks of speeds, they modify their response strategies differently to adapt to the particular conditions of their task. Two groups were created by presenting to one of them a fast range of objects and to the other one a slow range of objects. The three slowest objects of the fast-range group were the same as the three fastest objects of the slow-range group, to allow comparison. In an adaptive model employing rate-of-expansion that uses "any available cues" [3] to improve its performance, the prediction would be that speed effects only appear in the fast-range group where these speeds are, in fact, the slowest ones. They also found significant training effects.

The task included a pendulum which had to be released 400-ms before collision. This was simulated by moving the teaching input forward three time steps. We created two groups with seven speeds in each group, identically to Smith et al. experiment 3, with both groups having three overlapping speeds.

Ten simulations were made for every group. First they were trained for 35 epochs with full input data as seen in simulation 1. Every group was trained for 5 sessions of 5 epochs each, to test for training effects. Both groups were tested for the overlapping speeds in sessions 1 and 5.

3.2. *Results*

We used the Kruskal-Wallis test to analyze for significant difference. We obtained a significant effect of group ($p < 0.001$), so the fast-range group had more anticipated responses than the slow range group. Also training factor was significant ($p = 0.001$), so the performance tuned with experience. And finally, the speed factor was significant ($p = 0.001$), so the slowest of the overlapping speeds were responded to earlier than the fastest.

---

[4] In the optical state space, the axes are visual angle and rate of expansion. So, different response strategies can be drawn as linear combinations of both primitives.

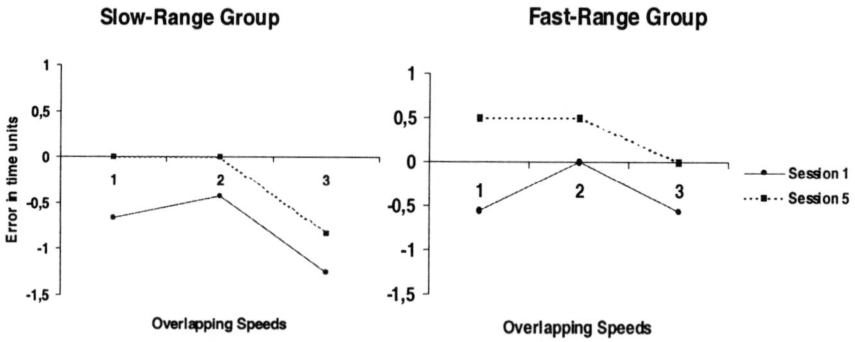

Figure 3. Results of simulation 2. The X-axis represents the three overlapping speeds and the Y-axis represents the performance of the SRNN: the correct responses were coded as o, anticipate responses were coded as 1 for each step of anticipation and late response were coded as -1 for each step of delay.

3.3. *Discussion*

While overall performance was replicated (speed, group and training effects), also the range manipulation produced the expected 'local minima' as found in the Smith et al. experiment in humans. The global training effect showed more tau-like responses as the groups were more trained. The conclusions of the Smith et al. article are applicable to this simulation, as subjects of the experiment as the SRNN use the available cues of the environment to improve their performance, in this case tuning their optical margin to the range constrains. Also, as a consequence of this dynamic system, the SRNN and the humans learn to adjust the use of the optical primitive (visual angle and its range of expansion) to the least-error strategy (Tau) from the initial strategy (Rho). Thus, the effects of anticipation produced by large sizes and slow speeds (typical of a strategy based on rate of expansion) tend to disappear with the training.

4. Conclusions

There are several models that try to explain TTC mathematically. On the other hand, there is agreement about several sources of information that are used to compute TTC and about the dynamic performance of humans when solving a TTC task. However, the mathematical models fail to explain this complex pattern of results unless the free parameters are fixed at different values for different tasks. Of course, training effects could be explained in terms of parametric changes. But a more parsimonious solution is available.

We have proposed a connectionist model that allows ontogenic analysis and explains adaptation and tuning to tasks and environment constraints. We show in simulations 1 and 2 that our SRNN learns the TTC task with a performance very close to human performance. Moreover, the network has the same pattern of errors as humans: largest and slowest objects were responded to earlier than others, which is consistent with human performance [4, 6-8, 14-15].

Another similarity between the SRNN and human behaviour is the improvement in the results with more training. Concretely, the network and human subjects use a "Rho-like" strategy in the beginning of their training, that implicates the size and speed effects, and their behaviour changes to a "Tau-like" strategy with more training, that implicates a reduction of these effects. Smith et al. use a linear parametric model that combines visual angle and rate of expansion. Their model allows to fit the relative weights of visual angle and rate of expansion; this way they can describe the progress from a rho-like strategy to a tau-like strategy employing parametric changes. Our model allows us to explain this transition of strategies guided by the data and without resorting to post hoc parametric adjustments. It is very difficult to imagine how an explanation based on a mathematical model without parametric changes (e.g. Tau or Rho) can account for this change of strategy. We can account for such a change by using a very simple connectionist model (only 101 units) without parametric changes.

A novelty introduced by our study is the possibility that the skill required for solving a TTC task could be learned. This point needs further investigation, but is an interesting possibility suggested from our model, because all preceding models based in mathematical models cannot explain how the subjects learn these functions or TTC solving at all.

Finally, as we don't code information about distance and speed we can say that the network uses an optic flow method, in opposition to a cognitive method. However, the SRNN didn't have texture information available. There are results that suggest that in the presence of texture information the humans use scale-change information to compute expansion rate more than optic flow [16]. It is an interesting way to continue our simulations.

Acknowledgments

We thank J. L. Luque for lending us his office, computers and coffee machine. Also we thank J. López-Moliner and P. L. Cobos for their useful comments. We thank specially two anonymous reviewers and their very meticulous work.

References

1. H. Sun and B. J. Frost. Computation of different optical variables of looming objects in pigeon nucleus rotundus neurons. *Nature Neurosci.* **1**, 296 (1998).
2. J. L. Elman. Finding structure in time. *Cognitive Science.* **14**, 179 (1993).
3. J. R. Tresilian. The revised Tau hypothesis: A consideration of Wann's analysis. *J. Exp. Psychol. Hum. Percept. Perform.* **23**, 1272 (1997).
4. R. W. McLeod and H. E. Ross. Optic-flow and cognitive factors in time-to-collision estimates. *Perception.* **12**, 417 (1983).
5. W. Schiff and M. Detwiler. Information used in judging impending collision. *Perception.* **8**, 647 (1979).
6. F. X. Li and M. Laurent. Occlusion rate of ball texture as a source of velocity information. *Percept. Motor Skills.* **81**, 871 (1995).
7. P. R. DeLucia. Pictorial and motion-based information for depth perception. *J. Exp. Psychol. Hum. Percept. Perform.* **17**, 738 (1991).
8. P. R. DeLucia and R. Warren. Pictorial and motion-based depth information during active control of self-motion: Size-arrival effects on collision avoidance. *J. Exp. Psychol. Hum. Percept. Perform.* **20**, 783 (1994).
9. D. N. Lee. A theory of visual control of braking based on information about time-to-collision. *Perception.* **5**, 437 (1976).
10. J. R. Tresilian. Empirical and theoretical issues in the perception of time-to-contact. *J. Exp. Psychol. Hum. Percept. Perform.* **17**, 865 (1991).
11. D. Regan and S. J. Hamstra. Dissociations of discrimination thresholds for time to contact and for rate of angular expansion. *Vision Res.* **33**, 447 (1993).
12. J. P. Wann. Anticipating arrival: Is the Tau margin a specious theory? *J. Exp. Psychol. Hum. Percept. Perform.* **22**, 1031 (1996).
13. N. Hatsopoulos, F. Gabbiani and G. Laurent. Elementary computation of object approach by a wide-field visual neuron. *Science.* **270**, 1000 (1995).
14. M. R. H. Smith, J. M. Flach, S. M. Dittman and T. Stanard. Monocular optical constraints on collision control. *J. Exp. Psychol. Hum. Percept. Perform.* **27**, 395 (2001).
15. J. L. Moliner and C. Bonnet. Speed of response initiation in a time-to-contact discrimination task reflects the use of η. *Vision. Res.* **42**, 2419 (2002).
16. P. R. Schrater, D. C. Knill and E. P. Simoncelli. Perceiving visual expansion without optic flow. *Nature.* **410**, 816 (2001).

Action and Navigation

APPLYING FORWARD MODELS TO SEQUENCE LEARNING: A CONNECTIONIST IMPLEMENTATION

DIONYSSIOS THEOFILOU
ARNAUD DESTREBECQZ
AXEL CLEEREMANS

Cognitive Science Research Unit, Université Libre de Bruxelles, CP191, Av. F.-D Roosevelt 50, Brussels B-1050, Belgium

The ability to process events in their temporal and sequential context is a fundamental skill made mandatory by constant interaction with a dynamic environment. Sequence learning studies have demonstrated that subjects exhibit detailed — and often implicit — sensitivity to the sequential structure of streams of stimuli. Current connectionist models of performance in the so-called Serial Reaction Time Task (SRT), however, fail to capture the fact that sequence learning can be based not only on sensitivity to the sequential associations between successive stimuli, but also on sensitivity to the associations between successive responses, and on the predictive relationships that exist between these sequences of responses and their effects in the environment. In this paper, we offer an initial exploration of an alternative architecture for sequence learning, based on the principles of Forward Models.

1. Introduction

1.1. *Sequence Learning*

Most aspects of cognition — consider for instance segmenting speech, riding a bicycle, planning your next day, apprehending music, reading - involve the ability to process sequences of events. Constant interaction with a dynamic, changing environment thus makes sensitivity to sequential structure a fundamental ability of cognitive agents. Laboratory studies of sequence learning have, over the past fifteen years or so, documented how participants can come to exhibit sensitivity to the sequential structure of streams of stimuli through, for instance, differences in their reaction time to items that are or are not predictable based on the temporal context in which they occurred. In such situations, participants are typically asked to react to each element of sequentially structured and typically visual sequences of events (e.g., Nissen & Bullemer [20]). Several versions of this basic paradigm can be distinguished. In *rule-based paradigms*, sequences either conform or fail to conform to an abstract rule that describes permissible transitions between successive stimuli. Rule-based paradigms can in turn involve either deterministic (e.g., Lewicki, Hill, & Bizot [18]) or probabilistic rules, as when the stimulus material is generated based on the output of finite-state grammars (e.g., Cleeremans [1]). By contrast, in the more common *simple repeating sequence paradigm*, a single sequence containing fixed regularities is repeated many times to produce a training set (e.g., Nissen & Bullemer, [20]). In this context, two issues continue to elicit debate. The first is to determine the exact nature of what is being learned in these situations. The second is to determine the extent to which sequence learning can occur implicitly, that is, without intention, and without verbalize-able knowledge of the acquired regularities. In this paper, we focus on the first issue: What type of information is learned in sequence learning? A good starting point to think about these issues is the Simple Recurrent Network (SRN) introduced by Elman [9], which we briefly describe in the following section.

1.2. *The SRN model of sequence learning*

The SRN network (Figure 1) uses back-propagation to learn to predict the next element of a sequence based only on the current element and on a representation of the temporal context that the network has developed. To do so, it uses information provided by so-called context units which, on every step, contain a copy of the network's hidden unit activation vector from the previous time step.

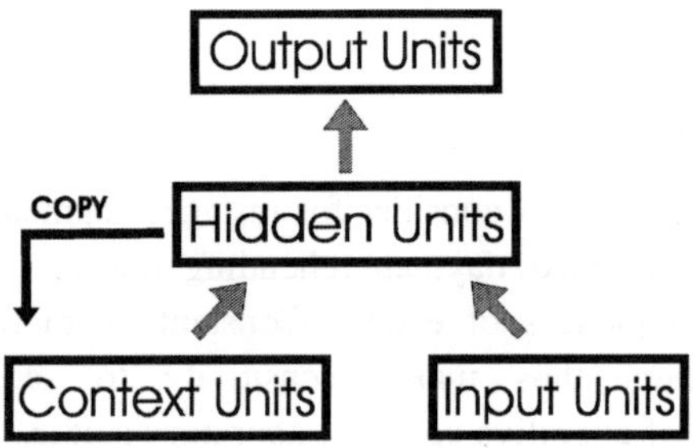

Figure 1. The SRN Network. Hidden Units are copied into Context Units on every step.

Previous work [1] [3] [4] has shown that the SRN is able to account for about 80% of the variance in sequential choice reaction time data. The SRN suffers from a number of limitations as a general model of sequence learning, however. Consider for instance a piano player. It only takes a moment to realize that several parallel sequences of events are unfolding concurrently in such a situation: Not only is the player processing a sequence of visual events as he reads the notes on the sheet of music, but he is also concurrently (1) producing a sequence or responses (the successive keystrokes) and (2) experiencing the consequences of his actions, that is a sequence of auditory tones. In other words, three sequences of events are involved: A sequence of stimuli S (the printed notes), a sequence of responses R (the keystrokes), and a sequence of effects E (the tones). Such a setting therefore provides opportunities to learn not only about the sequential relationships between successive stimuli (SS learning), but also about associations between successive responses (RR learning) and possible associations related also to the effects (RE learning).

Most theories have so far assumed that sequence learning only involves stimulus to stimulus (SS) relations, that is, that the system learns to anticipate future stimuli based on the current stimulus and on the temporal context in which it occurs [4], [16]. Other theories assume that it is response to response (RR) associations that are learned [19], or a combination of both SS and RR relationships [10]. Further, as each response always follows the presentation of a new stimulus, Ziessler [26] has proposed that subjects might also learn to predict the appearance of each stimulus as an *effect* of their previous response. In this sense, participants thus learn response–effect associations. Moreover, researchers like Hommel [15] have studied learning of stimulus – response – effect (SRE) associations, where subjects' responses are followed by an additional effect (usually a tone). Hommel showed that the presence of an effect facilitates responses independently of the stimulus. Finally, an additional SRE study [14] also demonstrated that even when the effects are irrelevant to the next stimulus, they still enhance reaction times. This enhancement is further proportional to the time that lapses between response and effect. Importantly, this study also suggests that responses in such situations are influenced by *anticipation of their effects*. Thus, sequence learning cannot be considered to be exclusively stimulus or response - based. Multiple learning systems instead contribute to performance (see [5] for a review). The implication of different neural circuits is also supported based on neuro - imaging evidence (see [25] for example). Addressing these issues from the point of view of developing relevant computational models therefore requires such models to distinguish between the respective contributions of perception, action, and memory to performance, something that the SRN is ill-equipped to deal with because it simply fails to distinguish between stimuli and responses at the level of its output units. In the next section, we explore a different class of networks (the forward models) in which the distinction of different input modalities, responses and anticipated stimuli is a feature of the architecture itself.

2. Forward Models

Forward Models, introduced in the connectionist literature by Jordan and Rumelhart [17], (but see also relevant in [11],[22]) and are aimed at solving the following problem: In many control systems, the actions that need to be performed so as to realize some goal cannot be reinforced directly. To see this, consider how your brain learned to issue the correct motor commands to the muscles of your arms when you first mastered the ability to reach for and grasp objects. Nothing in the world directly indicated to your motor cortex what the relevant patterns of activation were so as to make your arm move in the desired position. Rather, the only feedback that is available is based on comparisons between representations of the *target* perceptual state (your arm grasping the object) and of the *actual* perceptual state (the current position of your arm). Forward models make it possible to use this indirect feedback so as to learn the appropriate actions. Thus, the goal state and the current state are provided as inputs to the system, which then learns two different things on each interaction with its environment: (1) to predict the consequences of different actions, (2) to select actions appropriate to attaining its goals. To achieve this, FMs typically involve two interacting modules, as shown in Figure 2:

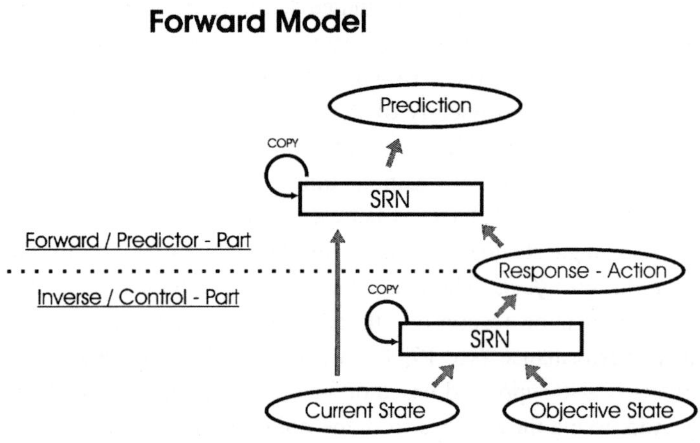

Figure 2. A general Forward Model includes two interacting modules. The inverse/control module produces actions based on the current state and the goal state. The forward/predictor module predicts the next state, that is, the consequences of carrying out the action in the environment. The model can learn proper actions based on desired future-environmental states or future anticipation. Considering sequence learning, RT depends not purely on the input, but on multiple Stimulus – Action – Effect relations that may exist. Therefore, a Forward Model is an ideal tool in connectionism to model such multi-dimensional dependencies.

The first module — the "inverse" or "control" part — takes the current state and the goal state as inputs. Based on that, it produces a response - an action that will influence the environment. The second module, which is called the "forward" or "predictor" part, receives both the current state and the current action as inputs and learns to predict the consequences of the action. Forward models are interesting not only from the perspective of learning how to control a system, but also from the perspective of understanding the relationships between perception and action. As a case in point, such models correspond almost exactly to the premises of the "enactive view" recently developed by Noë and O'Regan [21] (see also Varela [24]), which takes as a starting point that perception and action, far from constituting the terminal points of a purely feed-forward system going from one to the other, instead interact constantly.

3. Training in Forward Models

Simulation and training phases overlap: The model learns constantly. Training is always performed in two sequential cycles, one for the forward/predictor module of the model and one for the inverse/control module (Figure 2). In the first cycle, the current state and the objective state are presented to the network. Activity propagates through the first module - the control module - and generates an action on the output units of this module. At the very beginning of training the output value (represented action) will be just random. Based on this arbitrary response, we simulate the environment to find out what consequence this action will have. Then, we use this consequence as a target to train only the forward/predictor module of the model. In this way, after several cycles, the forward/predictor module comes to evaluate/predict how the environment will respond to the actions executed by the model.

In a second cycle, we use the same input used in the first cycle but instead of using the environment's simulated response as target pattern of the forward/predictor output, we use the goal that has been set - also as input - for the objective state. Then, we 'freeze' the weights of the forward/predictor module of the model so that only the weights of the control module will be adapted and we back-propagate the error based on the difference between our objective output pattern and the actual output. This way we force the control module to provide, after several training cycles, correct actions that will bring the future environmental state close to the desired objective, while we leave the forward/predictor module (which has to evaluate the environment) intact.

4. A forward model of sequence learning

How can we apply Forward Models to sequence learning? As the weakness of the SRN being one dimensional can not justify for a possible multi-modal contribution in sequence learning, we have examined the use of the Forward Model as a possible representative connectionist scheme for SRT tasks. Forward models can build modular architectures with multiple inverse/control and predictor modules as explained in [12], [25] and are capable to model actions as well as to predict effects of these actions in the environment. In this work we have implemented a single pair of an action – predictor system, and we show how it can be used to account for anticipation in SRT tasks, by modeling the subject's reaction time and the *Response to Stimulus Interval* (RSI) *

To model the SRT task, we have used the following variation of FM, as shown in Figure 3. The input that the inverse/control section receives at the beginning represents the next stimulus (location) that appears on the screen of the participant during the experiment. The participant then provides a response, pressing the corresponding key for this stimulus, that is the Action output value of the inverse/control section. As the participant repeats the block several times, he learns to predict the next value in the sequence. This prediction is represented by the output value from the forward/predictor part of the model which is then sent as a feed-back input to the inverse/control section of the network before the next stimulus arrives. If the predicted value is correct, it facilitates a faster response, as the new stimulus is congruent with network's prediction.

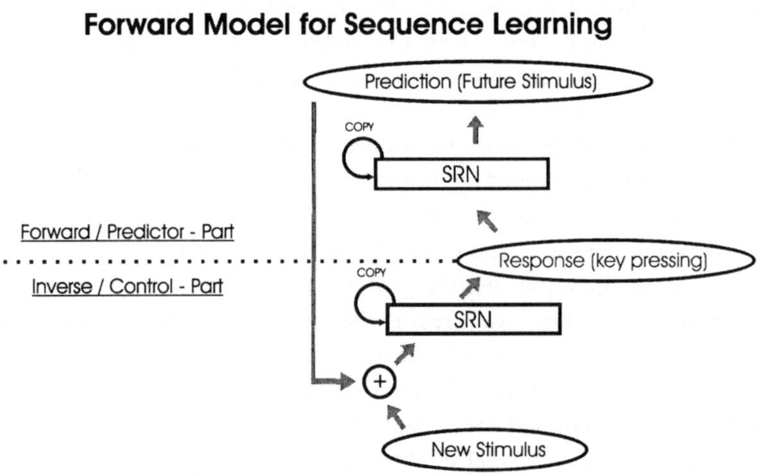

Figure 3. By feedback - cascading the future prediction of the next stimulus as estimated by the forward/predictor module of the model, we can model various RSI values and the effect of anticipation on RT in an SRT task: Participant selects a choice that in our model is represented by the Action response of the network. Then the predicted future stimulus is cascaded with a feedback and biases the next future choice of the participant (We cascade the predicted value on a number of cycles analogous to the RSI and the final accumulated value is added with the new input stimulus). This way, a large RSI value means more cascaded cycles from the predictor and better (faster) RT when the next input signal arrives in agreement with the empirical data.

* RSI - Response to Stimulus Interval: The amount of time that elapses between the response and the onset of the next stimulus during an experiment.

To represent response time (RT) and the RSI, we do not propagate the activation in the input all at once. Instead, we use a cascade scheme similar to [6] where the activation values are propagated over a sequence of cascade-cycles and input increases linearly until a threshold† is reached. If the forward/predictor module of the model has made a correct prediction, then when the new stimulus is blended to the predicted, it will result to a high activation value that will take only a few cascade cycles to accumulate and reach the required threshold. This small number of cascade cycles represents a small RT for the subject. Contrary, when a wrong prediction is blended with the new stimulus that arrives, since that the two maximum values of their two vectors do not coincide, it will require the accumulation of several cascade cycles until the required threshold is finally reached and produce a response. Consequently, response that will be provided by the inverse/control module of the network after several cascade cycles will represent the longer RT from the participant during the SRT task experiment

4.1. *Manipulating the RSI*

To explore the capacity of our model to account for sequence learning, we turned to a recent experiment carried out in our laboratory in which we studied the effects of manipulating the temporal variable [8]. More precisely, we explored how changes in the response-to-stimulus interval in the SRT task influenced sequence learning. In our experiments, the RSI could take three different values (RSI=0, 250, and 1500 ms). Experimental data (see Figure 5) have shown that reaction times tend to decrease over 15 blocks of practice but increase dramatically when the repeating sequence is switched to a transition sequence of the SRT task. This pattern of results was observed whatever the value of the RSI. Moreover, participants responded reliably faster when the RSI increased. Direct measures of learning, further revealed that higher value of the RSI tends to improve *explicit* sequence learning (see [8]). Here, we will simply focus, as a first step, on the ability of the forward model to account for the reaction time differences between the three RSI conditions in the SRT task.

To enable the model to represent RSI timing during training, we cascaded the activation transmitted from the output units of the prediction module to the input units of the inverse/control module in the model. The number of allowed back-propagated cycles represents different RSI values. For the highest RSI value (RSI1 in Figure 4), all cascade cycles were propagated and consequently 100% of the activation was transmitted and accumulated to the input units. For smaller RSI values, fewer cycles, result to 80% and 60% of the activation for the

† This threshold is based on the Luce Ratio of the layer.

intermediate (RSI2) and lower (RSI3) values of the RSI respectively. The more cycles transmitted, the higher the accumulated activation in the input, and the faster the network reaches threshold and provides a response when new stimulus arrives. Thus, the result is a faster RT. In contrast, a small number of feed-back transmitted cascade cycles, will result in lower accumulated activation value in the input, lower blended value when stimulus arrives and more necessary cycles to reach the threshold and provide a response. As an effect, the network will have a longer RT as is the case in the empirical data.

Figure 4 shows simulated reaction times for the three conditions. The figure shows that the RSI manipulation influences the model's performance in the same way as for human participants. RTs are smaller when the RSI increases. RT decreases with practice in all three conditions but increases when the sequence is switched to a different one in the thirteenth block. The simulated transfer effect is lower when no RSI is used as compared to the other two conditions. More research is needed to assess whether this discrepancy between experimental (in which transfer was equivalent in all three conditions) and simulated results is due to the set of parameters we have used in our simulations. Further simulations of performance in objective tests such as generation and recognition tasks [8] will be performed.

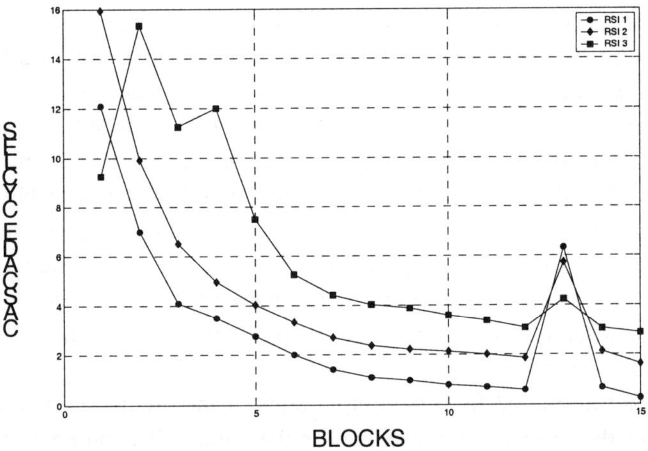

Figure 4. Simulation results – Average number of Cascade-Cycles / Block, required for crossing the output threshold at the inverse/control module of the Network for a response. Three different RSI times are shown. A large RSI value (RSI 1) represents longer time of anticipation. In the network model, this allows more cascade cycles to propagate from the forward/predictor module feed-back and influence the inverse/control part. As a result, the inverse/control module is ready when the new stimulus arrives and RT is small (only a couple of cascade cycles) as is the case with the empirical data from the subjects.

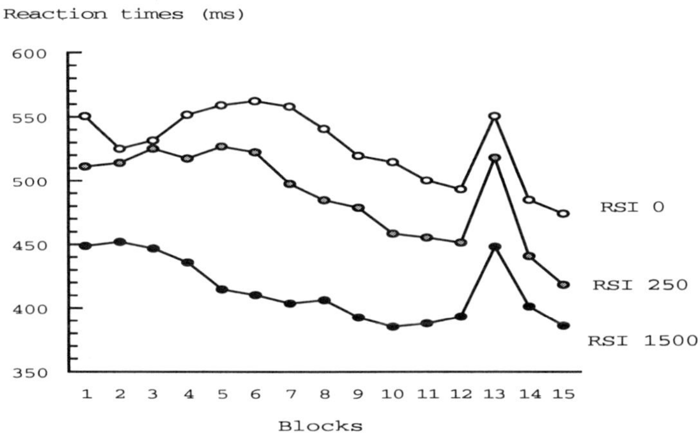

Figure 5. Empirical Data - Larger RSI values generate faster RTs from the participants. (See [8] for details).

5. Conclusions

Forward models appear to be a useful framework in which to think about sequence learning, particularly in the context of recent research that indicates that learning involves the simultaneous mastery of multiple modalities of sequential information. More importantly perhaps, forward models elegantly capture the fact that action is continuously influenced by the anticipated consequences of its effects (see Hesslow, [13]). In this respect, forward models represent a straightforward implementation of recent ideas about the relationships between action, memory, and perception, as embodied for instance in Noë and O'Regan's enactive view [21] or in Hoffmann's ideo-motor principle [14]. In this paper, we have only begun to explore how forward models can be applied to sequence learning. The model as it stands now is incomplete and limited in different ways, but it represents a promising first start towards an architecture for sequence learning that makes it possible to address a much wider range of empirical data than is currently possible with the SRN.

6. Acknowledgments

This work was supported by the European Commission Research Training Network Project: HPRN–CT–1999– 00065 and by institutional grants from the Université Libre de Bruxelles to the Cognitive Science Research Unit. Axel Cleeremans is a Senior Research Associate of the National Fund for Scientific Research (N.F.S.R., Belgium). Arnaud Destrebecqz is a scientific research worker within the same institution.

References

1. A. Cleeremans, *Mechanisms of Implicit Learning: Connectionist Models of Sequence Processing*, MIT Press (1993).
2. A. Cleeremans, A. Destrebecqz and M. Boyer, "Implicit learning: News from the front." *Trends Cognit. Sci.* **2**, 406-416 (1998).
3. A. Cleeremans and L. Jimenez, "Implicit sequence learning: the truth is in the details" *Handbook of Implicit Learning*, Staedler, M.A. and French P. (Eds.) 323-364, Sage Publications (1998).
4. A. Cleeremans and J. L. McClelland, "Learning the structure of event sequences" *J. Exp. Psychol. Gen.* **120**, 235-253 (1991).
5. B.A Clegg, G.J DiGirolamo and S. W. Keele, "Sequence learning" *Trends Cognit. Sci,* **2,** 275-281 (1998).
6. J.D. Cohen, K.Dunbar and J.McClelland, "On the Control of Automatic Processes: A Parallel Distributed Processing Account of the Stroop Effect" *American Psychol. Assoc.* **9**, 332-359 (1990).
7. A. Destrebecqz and A. Cleeremans, "Can sequence learning be implicit ? New evidence with the Process Dissociation Procedure" *Psychonomic Bull. Rev.* **8**, 343-350 (2001).
8. A. Destrebecqz and A. Cleeremans, "Temporal effects in sequence learning." *Attention and Implicit Learning*, Jiménez, L. (Ed.), John Benjamins, 181-213 (2003).
9. J. L. Elman, "Finding Structure In Time" *Cognit. Sci.* **14**, 179-211 (1990).
10. D.W. Fendrich, A.F. Hearly and L.E. Bourne, "Long-term representation effects for motoric and perceptual procedures." *J. Exp. Psychol. Learn. Mem. Cognit.* **17,** 137-151 (1991).
11. R. Grush, "The emulation theory of representation: motor control, imagery, and perception." *Behav. Brain Sci,* In press (2003).
12. M. Haruno, D.M. Wolpert and M. Kawato, "MOSAIC Model for Sensorimotor Learning and Control" *Neural Computation,* **13**, 2201-2220 (2001).
13. G. Hesslow, "Conscious thought as simulation of behaviour and perception" *Trends Cognit. Sci,* **6**, 242-247 (2002).
14. J. Hoffmann, A. Sebald, and C. Stöcker, "Irrelevant response effects improve serial learning in serial reaction time tasks" *J. Exp. Psychol. Learn. Mem. Cognit.* **27**, 470-482 (2001).
15. B. Hommel, "The cognitive representation of action: Automatic integration of perceived action effects" *Psychol. Res,* **59,** 176-186 (1996).
16. J.H. Howard, S.A. Mutter and D.V. Howard, "Serial pattern learning by event observation" *J. Exp. Psychol. Learn. Mem. Cognit.* **18**, 1029-1039 (1992).
17. M. L. Jordan, and D.E. Rumelhart, "Forward Models: Supervised learning with a distal teacher" *Cognit. Sci.* **16**, 307-354 (1992).

18. P. Lewicki, T. Hill and E. Bizot, "Acquisition of procedural knowledge about a pattern of stimuli that cannot be articulated" *Cognit. Psychol,* **20**, 24-37 (1988).
19. D. Nankemper and W.Prinz, "Stimulus and response anticipation in a serial reaction task." *Psychol. Res,* **60,** 98-112 (1997).
20. M. J. Nissen and P. Bullemer, "Attentional requirement of learning: Evidence from performance measures" *Cognit. Psychol.* **19**, 1-32 (1987).
21. K. O'Regan and A. Noe, "A sensorimotor account of vision and visual consciousness" *Behav. Brain Sci,* **24-5**, 939-973 (2001).
22. D.E. Rumelhart and J.L. McClelland (Eds.) *Parallel distributed processing: explorations in the microstructure of cognition*, Chap. 14, Vol.2, MIT press, 7-57 (1986)
23. A. Stock, and J. Hoffmann, "Intentional fixation of behavioral learning or how R-E learning blocks S-R learning." *European Journal of Cognitive Psychology*, In Press (2002)
24. F.J. Varela, E. Thompson and E. Rosch, *The Embodied Mind: Cognitive Science and Human Experience*, MIT Press, (1991).
25. D. M. Wolpert and M. Kawato,"Multiple paired forward and inverse models for motor control." *Neural Networks* **11**, 1317-1329 (1998).
26. M. Ziessler, "Response-effect learning as a major component of implicit serial learning" *J. Exp. Psychol. Learn. Mem. Cognit.* **24**, 962-978 (1998).

THE SIMULATION OF CHARACTER PRODUCTION BEHAVIOURS IN CONNECTIONIST NETWORKS

F. M. RICHARDSON, N. DAVEY, L. PETERS AND D. J. DONE

Departments of Computer Science and Psychology,
University of Hertfordshire, College Lane, Hatfield, Herts, AL10 9AB, England
E-mail: F.1.Richardson, N.Davey, L.Peters, D.J.Done@herts.ac.uk

In order to draw (or copy) a character, a process of *linearising* takes place. In this process the complete static form of the character is broken down into a temporal sequence of strokes for graphic production. According to Thomassen, Meulenbroek and Tibosh [1], individuals develop their own production rule base, which is reflected as tendencies or strategies for graphic production. Occasionally, these principles of production come into conflict resulting in a variable sequence of production for some characters. The work described in this paper uses a connectionist modeling approach to investigate the emergence of production-based behaviours in the sequential production of characters [2]. Here, the emergence of human-like production behaviours is simulated, without the need for explicitly imposed heuristics. Demonstrating that not only are connectionist networks capable of emulating the production-sequence behaviour of humans, but also that rule-like tendencies can emerge naturally upon the basis of learning.

1. Introduction

Handwriting involves the structured sequencing of movements in order to produce a letter or a string of letters. Generally, letters are produced with regularity according to basic production principles, which are formed through a variety of environmental influences [3], [4], [5]. These may include culture and handedness, in combination with the level of experience and practice [1], [6], [7], [8]. In order to draw or copy a character, a process of *linearising* takes place [1]. In this process the complete static letterform is broken down into a temporal sequence of strokes for production.

Often the sequence in which strokes are produced is influenced by the direction in which the script will be read [9]. In the case of writing that is produced from left to right, the sequence of strokes usually commences at the leftmost point of the letter, and then progresses through neighbouring strokes. However, in some cases characters are produced starting with the topmost point or the first vertical line of the character. The selection of the appropriate production sequence is one that is usually the most economical in terms of movement. This variety of movement trajectory usually reduces the opportunity for production errors to occur, although the demand for accuracy occasionally leads to the implementation of specific strategies to reduce error.

A further means of reducing the potential for errors to occur during the production process is consistency. For example, the curved segment/s of letters P, R, B, and D are all produced in a clockwise direction. In the case of letters R and B, this type of movement also preserves economy, in that the following stroke or segment continues from the previous, without the need for any positional change, thus economy and consistency have a close relationship. The overall consistency in letter production is reflected in graphic production models used to teach basic handwriting [9], [10], [11]. These models are very similar and usually contain style as opposed to production differences [12]. This consistency allows rules for production to be extracted from these models, such as those described earlier.

However, rule-based descriptions are not complete descriptions of character production processes. For some characters production rules can come into conflict. Experimentation has shown that letters with conflicting production processes take longer to be produced and have less fluent and more variable movement trajectories, resulting in different production sequences for the same letters [1]. For example, a letter T' can be produced starting with either the horizontal crossbar or the vertical stroke [12]. Further evidence of this conflict comes from a comparison of graphic production models, which has found differences in the production processes for specific characters between models [12]. Thus, production should not be considered to be completely stable and absolute, but more as a series of probabilistic rules or tendencies with varying strengths [1].

2. Simulating Character Production Sequences

As the process of character production has traditionally been described in terms of a series of rules, it could be considered that simulating character production behaviour is somewhat incompatible with a connectionist approach. However, as highlighted in the previous section, rule-based descriptions do not offer a complete description of character production processes, nor do they account for the emergence of production behaviours as a result of experience and practice as in humans. In this respect the characteristic of connectionist models to produce behaviour as a result of experience makes them well suited to the examination of the emergence of production behaviours. The added advantage of using connectionist networks is that it is possible to directly manipulate the nature and level of their experiences (something not possible with human participants). This allows a structured investigation into character production behaviour to take place with the aim of providing an emergent account of both rule-like and probabilistic production tendencies.

2.1. *The Task*

The simulations conducted dealt specifically with the issue of stroke sequences in character production. The task of the networks upon presentation of a static character at its input array was to identify and output the appropriate sequence of strokes used to produce the character. A network carrying out this task can be seen in Figure 1.

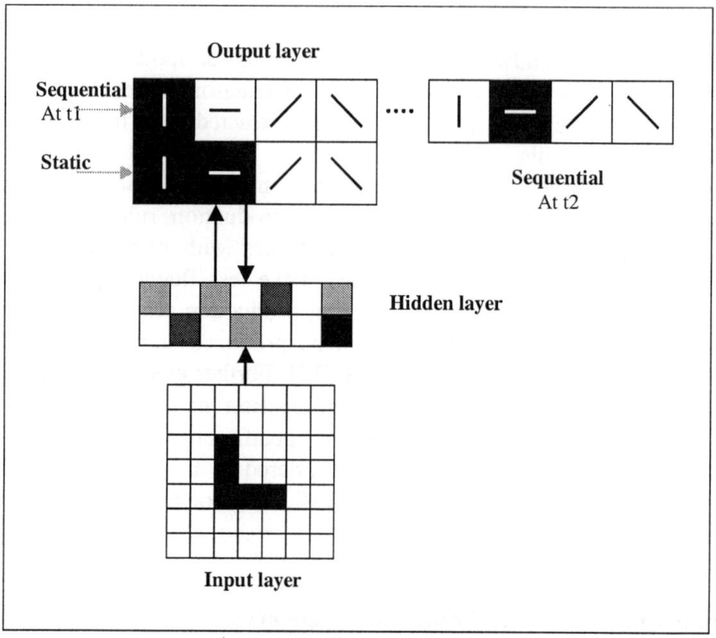

Figure 1. Shows a network performing the sequential production task. Active units are displayed in black, with arrows indicating full connectivity between layers. The network is presented with a complete static character as input. As output the network is required to identify the component line-types of the character (as displayed in the static section of the output layer) and also output the sequence of strokes which should be used to produce the character (on the sequential section of the output layer).

2.2. *Training*

Networks were trained on the task using back-propagation though time [13]. A total of fifteen runs were conducted in order to establish the general trend in behaviour for these networks. Each run was initialised with a different set of starting weights. When presented with an input character, networks were required to give an output response irrespective of the size or location of the character on the input array. All characters were composed of two different line-types of varying combinations (vertical |, horizontal _, diagonal-obtuse /, diagonal-acute \), examples of which can be seen in Figure 2. All networks were trained to their point of maximal generalisation performance. In order to explore the emergence of the production behaviours exhibited by the networks in detail, two separate phases of training were conducted, during which the learning experiences of the network were developed. The following subsections describe the motivation for this training and the expected behaviour of the networks.

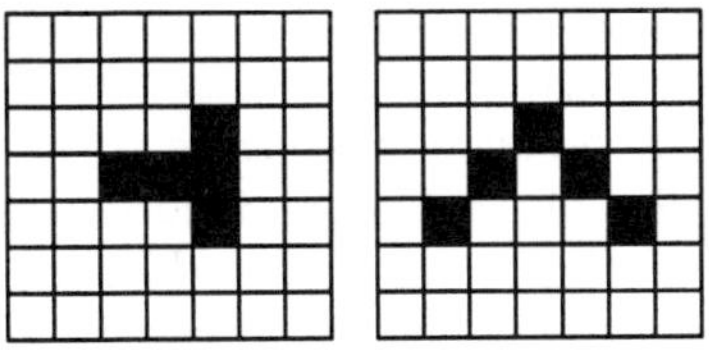

Figure 2. Shows two examples of the characters presented to the network during training.

2.2.1. *Phase One*

During *phase one* of training, networks learnt to produce a total of eight characters in a manner consistent with a left-to-right production sequence. At this point in learning it was expected that during testing the networks would produce generalisation behaviour consistent with this - producing all characters using a left-to-right production sequence

2.2.2. *Phase Two*

In this phase networks were given an additional increment of training during which they learnt about characters for which only top-to-bottom was a viable production sequence. It was expected that networks during this phase would produce a variable production behaviour for test characters where either a left-to-right or top-to-bottom production sequence was equally viable, but would result in a different production sequence in either case.

2.3. *Testing*

Testing was carried out upon completion of each phase of training. Networks were assessed in terms of their ability to produce the appropriate production sequence for a series of completely novel characters. These testing characters were displayed in a variety of size and location combinations on the input array.

2.3.1. *Phase One*

In this phase production behaviour was tested using three sets of novel patterns known as *(a) normal, (b) ambiguous* and *(c) exceptions,* which can be seen in Figure 3. The set of *normal* patterns consisted of four characters that conformed to a left-to-right production sequence. It was expected that the network would be able to produce all of these characters correctly. The *ambiguous* set consisted of two characters for which the production sequence is known to vary in humans. For example, a comparison of different production models for capital letters of the alphabet used to teach children found that the letter 'T' has two viable methods of production producing either the horizontal cross-bar or the vertical stroke first [12]. It was expected that networks from this phase would produce characters in a left-to-right production sequence, as all characters on which the network was trained at this phase conformed to this production heuristic. The *exceptions* set consisted of two characters for which the appropriate sequence of production conforms to a top-to-bottom sequence only. It was expected that networks from this phase would not be able to produce the appropriate output sequence for these characters, as applying a left-to-right heuristic does not give a clear sequence of production.

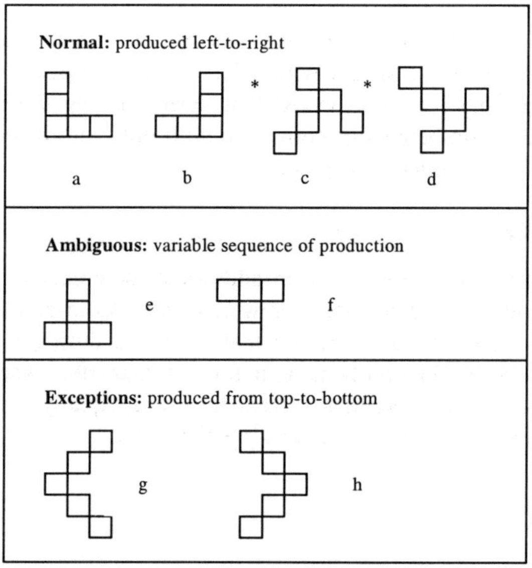

Figure 3. Shows the three sets of characters used during testing, each labelled with the applicable production heuristic. For *normal* characters shown with a *, the actual sequence of production varies if a different heuristic is applied. For the *ambiguous* set, characters could be produced according to either production heuristic.

2.3.2. *Phase Two*

In this phase the production behaviour of the network was tested for *normal* and *ambiguous* characters. This was in order to determine what production behaviour would emerge as a result of introducing the top-to-bottom heuristic. It was expected that the production behaviour for the *normal* characters would remain unchanged, but that for *ambiguous* characters would vary as a result of competition between the left-to-right and top-to-bottom production heuristics of which the network had knowledge at this point in testing.

3. Results

All network runs were assessed in both phases in terms of their output responses in order to determine the production sequences generated for each character. These outputs were categorised as to whether they were a *definite output* or a *no-response*. An output was considered a *definite output,* if the activation of the target units were 0.50 or over, and no non-target units had an activation of 0.50 or higher at either time-step. Any other outputs were classified as a *no-response.* For the *normal* and *exceptions* test sets, *definite outputs* were compared with the appropriate output response, in order to determine their accuracy. For the *ambiguous* test set, output responses were assessed simply in terms of the sequence of production, as either of the two possible sequences of response was considered to be viable.

3.1. *Phase One*

Firstly, it should be noted that all network runs were able to produce a large number of definite outputs during testing. Generally, output sequences from the networks for *normal* characters were as expected, being in accordance with a left-to-right production sequence. Interestingly, for *ambiguous* characters, networks produced a variety of output sequences, on some occasions producing characters in a left-to-right sequence, and for others in a top-to-bottom sequence. This behaviour was unexpected, but indicated that networks had also incorporated knowledge of *top-to-bottom* production sequences during learning. Support for this view comes from the mixed output responses for *exceptions,* for which networks were unable to consistently produce the appropriate output sequence. This general trend in behaviour over all runs is shown in Figure 4. In order to illustrate this behaviour in more detail in terms of individual responses, a summary of the output sequences obtained for each test character over five network runs is shown in Table 1.

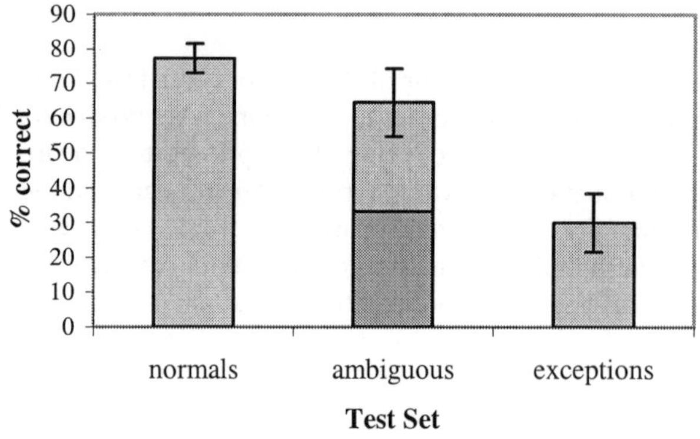

Figure 4. Shows the mean percentage of correct outputs for each test set over all network runs for phase one. Correct outputs for the *ambiguous* set are shown for HV (bottom) and VH (top) ordered combinations. It can be seen that networks were able to produce a large number of correct/appropriate output responses for *normal* and *ambiguous* characters, but struggled with *exceptions.*

Table 1. A summary of output responses for five network runs for the three different types of test sets upon completion of phase one of training. Letters correspond to the characters shown in Figure 3. Correct output responses for *normal* characters are indicated by '*', no-responses by '-', and incorrect by 'X'. For *ambiguous* characters, the response sequences of the networks are shown, for example, 'HV' indicating a horizontal followed by vertical line response sequence. An 'M' indicates responses that varied across trials. For *exceptions* varied responses are also indicated by an 'M', with correct, incorrect and no responses being indicated in the same manner as *normal* characters.

	Normals				*Ambiguous*		*Exceptions*	
	a	b	c	d	e	f	g	h
1	*	*	*	-	HV	HV	-	X
2	*	*	*	-	HV	VH	M	M
3	*	*	*	-	M	VH	M	M
4	*	*	*	*	VH	VH	M	M
5	*	*	*	*	VH	M	-	-

3.2. *Phase Two*

As noted for phase one, here too all network runs were able to produce a large number of definite outputs during testing. As in phase one, the majority of *normal* characters were produced in accordance with a left-to-right production sequence, with some networks showing a small improvement in output performance. As networks had learnt to produce *exception* characters during training in phase two, they were now able to appropriately apply the top-to-bottom sequence of production for these characters. The networks continued to produce a range of output response sequences for *ambiguous* characters, which varied both between and within network runs. Thus, following the second phase of training, in testing networks exhibited rule-like behaviour as well as varied production behaviour for characters with conflicting production heuristics, similar to that seen in humans (see section 1) [1], [12]. This general trend in behaviour over all runs is shown in Figure 5. A more detailed view of this behaviour can be seen in Table 2, which shows a summary of output responses over five network runs.

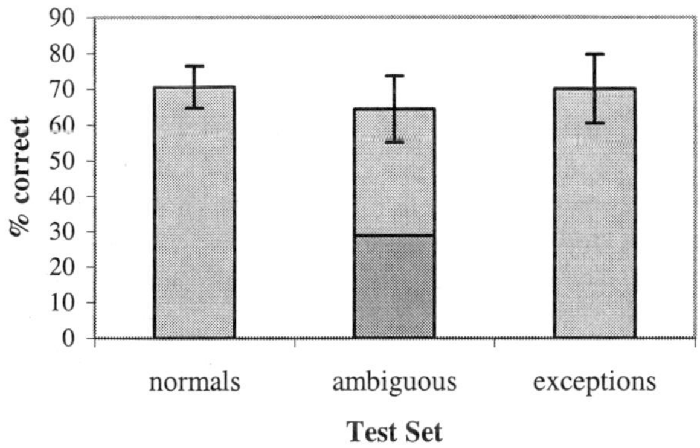

Figure 5. Shows the mean percentage of correct outputs for each test set over all network runs for phase two. Correct outputs for the *ambiguous* set are shown for HV (bottom) and VH (top) ordered combinations. It can be seen that networks were able to produce a large number of correct/appropriate output responses for all test sets, thus exhibiting a rule-like behaviour for *normal* and *exception* test characters, and varied output responses for *ambiguous* characters.

Table 2. A summary of output responses for five network runs for the three different types of test sets upon completion of phase two training. Letters correspond to the characters shown in Figure 3. Correct output responses for *normal* and *exception* characters are indicated by '*', no-responses by '-', and incorrect by 'X'. For *ambiguous* characters, the response sequences of the networks are shown, for example, 'HV' indicating a horizontal followed by vertical line response sequence. An 'M' indicates responses that varied across trials.

	Normals				*Ambiguous*		*Exceptions*	
	a	b	c	d	e	F	g	h
1	*	*	-	*	HV	HV	*	*
2	*	*	*	*	HV	M	*	*
3	*	*	*	-	M	M	*	*
4	*	*	*	*	VH	M	*	*
5	*	*	-	*	VH	M	*	*

4. Summary and Conclusions

The series of experiments reported in this paper were conducted in order to gain an insight into how experience influences the generation of character production sequences. Networks were required to use their knowledge about character production acquired through learning in order to be able to produce character production sequences for a series of completely novel testing characters. Networks underwent two phases of training, first learning about characters produced in accordance with a left-to-right production sequence, and then also learning about characters that could be produced using a top-to-bottom production sequence.

Testing completed at the end of each phase of training revealed that not only were the networks capable of generalising upon the basis of their learning experiences, but also exhibited some interesting production behaviours as a result. During phase one of testing it was found that the networks utilised their knowledge of the left-to-right production sequence in order to produce a series of *normal* test characters. Interestingly, upon testing with a series of *ambiguous* characters with conflicting production heuristics, networks produced a variety of production sequences. This indicated networks had also learnt about other production heuristics, and consequentially exhibited variant production behaviour for characters identified as having conflicting production heuristics [1], [12]. Following further training in phase two, during which networks also learnt about characters produced in a top-to-bottom sequence, the production behaviour of networks was extended to include characters produced either left-to-right, top-to-bottom, and varied production behaviour for characters with conflicting heuristics. Thus, exhibiting a full range of production behaviours emergent as a result of learning experience, without the need for explicitly imposed heuristics.

Generally, the model illustrates the exceptional generalisation ability of recurrent networks. Overall, within the context of character production this work is extremely interesting, as it demonstrates that connectionist models are capable of simulating rule-like and variant production-based behaviours simultaneously, and provides an emergent account of character production behaviour.

References

1. A. Thomassen, R. Meulenbroek and H. Tibosh. Latencies and kinematics reflect graphic production rules. *Human Movement Science* 10, 271-289 (1991).
2. F. M. Richardson, N. Davey, L. Peters, D. J. Done and S. H. Anthony. Connectionist models investigating the representations formed in the sequential generation of characters. *In Proceedings of the 10th European Symposium on Artificial Neural Networks. D-side publications, Belgium,* 83-88 (2002).
3. J. Alston and J. Taylor, *Handwriting: theory, research and practice.* New York: Nichols (1987).
4. R. Meulenbroek and A. Thomassen. Stroke-direction preferences in drawing and handwriting. *Human Movement Science* 10, 247-270 (1991).
5. M. Simner, C. Leedham and A. Thomassen (Eds.) *Handwriting and drawing research: basic and applied issues.* IOS Press (1996).
6. D. Desbiez, A. Vinter and R. Meulenbroek. Biomechanical and perceptual determinants of drawing angles. *Acta Psychologica.* 94, 253-271 (1996).
7. R. Meulenbroek and A. Thomassen. Exploitation of elasticity as a biomechanical property in the production of graphic stroke sequences. *Acta Psychologica* 82, 313-327 (1993).
8. J. Alston, and J. Taylor, *The handwriting file,* Cambridge: Living and Learning Ltd. (1988).
9. J. Alston, and J. Taylor, *Handwriting helpline,* Manchester: Dextral Books (1990).
10. S. Armstrong, *Handwriting.* London: The English and Media Centre (1994).
11. H. Hadley, *The handwriting book,* Cheltenham, England: Stanley Thornes Ltd (1996).
12. F. M. Richardson. Handwriting. TR. *University of Hertfordshire* (2000).
13. D. E. Rumelhart, G. E. Hinton and R. J. Williams. Learning internal representations by error propagation. *In Parallel Distributed Processing,* vol. 1, chap 8. Cambridge: MIT Press (1986).

AN INTEGRATION OF TWO CONTROL ARCHITECTURES OF ACTION SELECTION AND NAVIGATION INSPIRED BY NEURAL CIRCUITS IN THE VERTEBRATES: THE BASAL GANGLIA

B. GIRARD*, D. FILLIAT**, J.-A. MEYER*, A. BERTHOZ*** AND A. GUILLOT*

**AnimatLab/LIP6, CNRS-University of Paris 6,*
8 rue du Capitaine Scott,
75015 Paris, France
E-mail: {benoit.girard;jean-arcady.meyer;agnes.guillot}@lip6.fr

***DGA/Centre Technique d'Arcueil,*
16 bis av. Prieur de Côte d'Or
94110 Arcueil Cedex, France
E-mail: david.filliat@etca.fr

****Laboratoire de Physiologie de la Perception et de l'Action,*
CNRS-Collège de France,
11, place Marcelin Berthelot
75231 Paris Cedex 05, France
E-mail: alain.berthoz@college-de-france.fr

A control architecture of action selection, inspired by the neural loops of the dorsal part of the basal ganglia –subcortical nuclei of the vertebrate's brain– was proved to be able to solve a minimum survival task. The present paper concerns the connection of this architecture with a navigation system. This connection is inspired by recent hypotheses concerning the role of a ventral nucleus of the basal ganglia in integrating spatial, motivational and sensorimotor information. The ventral loop selects locomotion actions generated by various navigation strategies and modulated by motivations. The dorsal loop is in charge of non-spatial task selection and of coordination with the ventral loop. Implemented in a simulated robot performing the same survival task as in the previous experiment, the whole architecture improves the robot's survival thanks to map building and path planning abilities. Furthermore, the robot is also able to occasionally overlook the information recorded in its cognitive map in order to behave opportunistically, i.e. to reach an unexpected but visible resource, instead of a memorized but remote one. These results are discussed in terms of biological and robotic contributions.

1. Introduction

Contemporary robots are predominantly single-task systems, operating in specially-designed environments, in which they perform pre-programmed sequences of actions. While such systems have proved appropriate and useful on the factory floor, many future applications of robotics will require control systems with much greater flexibility. An effective strategy for designing flexible control systems is to reverse-engineer the biological control systems evolved in animals.

Among the essential functions for an autonomous system is the ability to decide which action should be done next to achieve its goals as well as the ability to orient itself in an environment in order to reach the various resources required by these goals [6]. Various computational models have proposed efficient mechanisms for these functions, respectively called action selection and navigation (see [16,3,14] for reviews). However, most action selection models use rudimentary navigation strategies while navigation models neglect the motivational choices which should guide the selection of their goals. To our knowledge, the integration of spatial information in a control architecture of action selection was not addressed as a specific issue.

Our objective is to design a bio-inspired architecture interfacing action selection and two navigation strategies of different complexity levels, taking inspiration from recent hypotheses about the role of the basal ganglia (BG) –a group of subcortical nuclei in the vertebrate's brain. These neural structures are assumed to be involved in action selection.

The BG are organised into segregated channels which receive inputs from the whole cortex and project to both brainstem nuclei and frontal parts of the cortex via thalamus nuclei, forming cortex-BG-thalamus-cortex loops. The selection occurs between the channels, a limited number of which being activated at any time. Each channel is supposed to be associated with one behaviour –a simple motor act like 'turn left' or an action pattern like 'feeding'. The BG select these channels by disinhibition [1]: the output nuclei of the basal ganglia are inhibitory and tonically active, therefore, in the absence of cortical input, they keep their targets –thalamus and brainstem nuclei– under constant inhibition. Specific patterns of activation in the cortex remove this inhibition in some channels and the related targets are consequently able to contribute to the behaviour.

There are three main cortex-BG-thalamus-cortex interconnected loops which are functionally distinct: motor (dorsal), associative (dorsal) and limbic (ventral). Low-level action selection using sensory-motor data is supposed to be located in the motor loop and more cognitive processes are assumed to be managed by the associative loop while the integration of motivation and topologic navigation involves the limbic loop [18]. In this specific loop, the BG input nucleus –the nucleus accumbens (NAcc)– receives spatial information provided by the hippocampus (localisation [15]) and the prefrontal cortex (path planning [17]). However, the coding of this information in the NAcc core is still not clearly deciphered [11].

In a previous work, we have tested a computational model of action selection inspired by the dorsal loop, designed by Gurney *et al.* [7,8], in a survival task. It was implemented on a Lego Mindstorms robot which had to select efficient sequences of actions in order to find vital resources and it was able to solve this action selection problem [5]. However this robot followed the most rudimentary navigation strategy to find these resources: a random walk. To improve the survival of an artificial system in a complex environment, our objective is to add to this architecture a second circuit –simulating the ventral loop– which would select locomotor actions according to more complex navigation strategies: *object approach* and *topological navigation*.

After having described the survival task and the characteristics of the simulated robot, we will introduce the details of our computational model. The results of two experiments focusing on the contribution of navigation to action selection issues will be discussed in terms of biological and robotic advances.

2. Survival task

The model is evaluated in the same survival task as in [5]. The simulated robot has to maintain above 0 a constantly decreasing internal variable called *Energy*. To do so, it has to find "ingesting" places where it can reload its *Potential Energy*, then find "digesting" places to turn its *Potential Energy* into *Energy*. *Potential Energy* and *Energy* fluctuate between 0 and 1, and the consumption of a full charge of *Energy* without reloading allows the robot to survive only 16 min.

3. Environment and robot

The experiments are performed in 2D environments, where the resources are represented by $50cm \times 50cm$ squares of specific colours. The robot is circular ($30cm$ diameter) and equipped with an omnidirectional linear camera providing every 10° the colour of the nearest object. A resource is considered available when it occupies more than 60° in the visual field (its centre is within a range of less than $70cm$ from the robot's centre).

The robot has three internal variables: *Energy*, *Potential Energy* and *Disorientation* (linked with navigation). For each type of resource –*Energy* and *Potential Energy*– three external variables (*Prox*, *maxProx* and *Avail*) are computed using the camera's image. *Prox* is a vector measuring the proximity of the resource in every direction (used for the *object approach*), *maxProx* is the maximum of this vector and *Avail* is a boolean which is true if the resource is available ($distance < 70cm$). The possible actions are: *Reloading Energy*, *Reloading Potential Energy* and *Moving in a given direction.*

4. Model

Both ventral and dorsal loops are simulated by the Gurney *et al.* model –named GPR according to the initials of the authors [7,8]. The GPR was also chosen to model the ventral loop because it is –anatomically and physiologically– very similar to the dorsal loop [12].

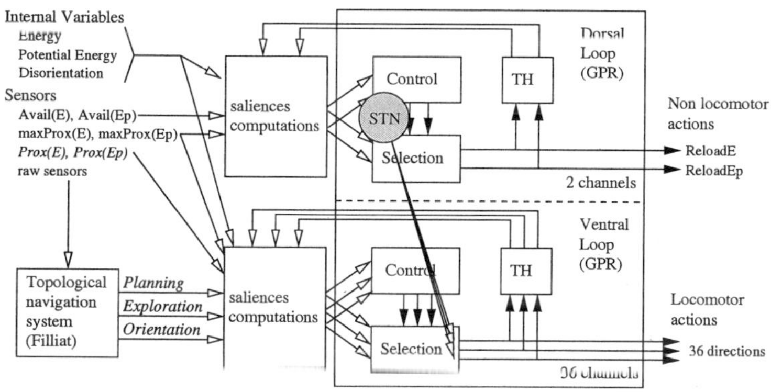

Figure 1. Integration model of navigation and action selection, calling upon two basal ganglia loops. The STN of the dorsal loop, which provides the interconnection with the ventral loop, is highlighted. GPR: Gurney *et al.* model; black arrows: inhibitions; white arrows: excitations; italics: 36 value vectors.

Only the main characteristics of the GPR will be summarized here (for details, see [7,8]). The GPR identifies two sub-circuits in the basal ganglia (Fig. 1, GPR boxes): a *selection circuit*, which processes the cortical input and generates the inhibitory output of the basal ganglia and a *control circuit*, which regulates the activity of the selection circuit and insures that inhibitions are always contrasted (disinhibition of the selected channel and strong inhibition of the others). The thalamo-cortical loop (TH) provides positive feedback to cortical inputs.

In the GPR, each channel is represented by one leaky-integrator neuron in every nucleus of the BG. The interconnections between nuclei are based on their neuro-anatomy, while the connection weights are fixed and derived from the connection densities. The *selection circuit* operates through two mechanisms: local recurrent inhibitory connections in the D1 striatum act as a "winner-takes-all" which disinhibits the winning channel, while the global "off-centre on-surround" structure of the network reinforces the inhibition of the losing channels and thus the discrepancies between the winning channel and the others. The *control circuit* is similar to the *selection circuit*, except that its output is directed to the *selection circuit*. This modulation of the *selection circuit* enhances its selectivity and allows for effective selection irrespective of the number of channels.

For every channel, the processed input is called *salience*, which represents the commitment towards displaying the action associated to that channel. It is computed by a sigma-pi neuron whose inputs are the external and internal variables as well as the feedback value provided by the thalamo-cortical loop, processed by transfer functions (linear, square and square root combinations). Because the system currently lacks any learning capacity, all the parameters of the transfer functions and the weights of the neurons are hand-tuned.

4.1. *Ventral loop*

Our main hypothesis is that the ventral loop selects directions of movement, given by simple sensor processing (for the *object approach* strategy) and by a navigation model (for the *topological navigation* strategy). The *object approach* generates movements induced by the proximity and direction of interesting in-sight resources. The *topological navigation* system (a modified version of the model of Filliat [2]) builds a topological map of the environment, using allothetic (vision and sonars) and idiothetic (self-movement) data. This map is made up of nodes representing locations in the environment. Links between nodes represent movements the robot has previously done between two locations. Within such a map, the robot's position is represented and calculated probabilistically according to algorithms inspired by hidden Markov models.

There are 36 channels, each of them coding for a 10° direction range.

Their saliences are influenced by vectors of 36 intensities computed by each strategy. The *object approach* is implemented by the two *Prox* vectors, in which directions leading to close resources are favoured. The *topological navigation* system produces three direction vectors: a *planning* vector, indicating the directions of the known resources, an *exploration* vector, favouring the directions leading to unmapped areas and an *orientation* vector, leading to well-known areas.

This *topological navigation* system puts some constraints on action selection because the robot is committed to regularly return to previoulsy mapped areas in order to ensure the building of a precise map. This need is expressed by a *Disorientation* variable produced by the system, which increases when the robot enters unexplored areas and decreases when it goes back in known areas. *Disorientation* is used as an additional internal variable in the computation of saliences.

At the output of the model, the selected direction is the result of a "soft-switching" process in which every channel disinhibited below a given threshold influences the final decision according to its level of disinhibition.

4.2. *Dorsal loop and loops interconnection*

The dorsal loop is kept to handle the non-locomotor behaviours. In our implementation, it corresponds to the reloading actions (reloading *Energy* and *Potential Energy*). As these two actions are not compatible, they are subject to "hard-switching": only the most disinhibited one is activated.

These reloading actions imply that the robot stops moving. Consequently, when the dorsal loop selects one of these actions, it has to be able to simultaneously influence the ventral loop in order to prevent locomotion. According to the so-called "trans-subthalamic pathway" hypothesis [10], this is modelled by excitatory projections from the part of the subthalamic nucleus (STN in Fig. 1) dedicated to one BG loop, to the output of the other loop. These additional excitations, generated by the selection of an action in the dorsal loop, and fed into the output nucleus of the ventral loop, increase its output inhibitions thus preventing the selection of any direction.

5. Experiments

Two experiments were carried out to respectively test the efficiency of the topological navigation integration and the capacity of the model to use its two navigation strategies in a situation requiring opportunistic decisions.

5.1. *Topological navigation efficiency*

In this experiment, the robot runs in the environment ($7m \times 9m$) depicted in Fig. 2 (left): one resource of each type is available, but it is impossible to see one resource from the vicinity of the other one. Consequently, a "reactive" robot following an *object approach* strategy only has to rely on random exploration to find hidden resources. On the contrary, after a first phase of random exploration and map building, a robot using a topological map should be able to reach desired resources.

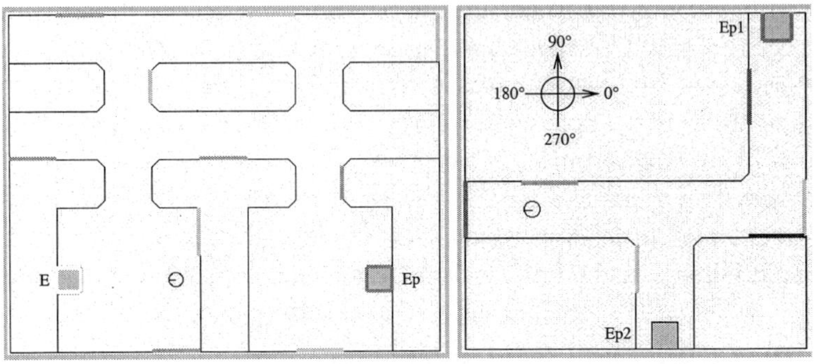

Figure 2. Test environments used in experiment 1 (left) and experiment 2 (right). E: "digesting" resource; Ep: "ingesting" resource.

Ten tests, with a four-hour duration limit, were run for two simulated robots, one with topological navigation capacities (robot A) the other without (robot B). *Energy* and *Potential Energy* are initially set to 1. Table 1 (left) shows that robot A is able to survive significantly longer ($p < 0.01$, Mann-Whitney test) than robot B.

Table 1. Experiment 1 (left): comparison (U Mann-Whitney test) of experiments duration medians of robots A and B. Experiment 2 (right): resources choice using three different configurations of weights for planning (Plan) and object approach (OA) vectors.

Durations (s)	**Median**	**Range**
Robot A	14431.5	2531:17274
Robot B	4908.0	2518:8831
U test	U = 15	$p < 0.01$

Weights		**Selection**	
Plan	**OA**	**Ep1**	**Ep2**
0.65	0.55	13	2
0.55	0.55	7	8
0.45	0.55	2	13

In [5], action selection was only constrained by the artificial metabolism. Here, the addition of the *topological navigation* system generates a new constraint: if the robot is too much disoriented, it has to come back to known areas in order to build a precise and useful map. Despite this additional constraint, the integration of the *topological navigation* system into the action selection system proves to be efficient in enhancing the life span of robots.

However, during three robot A tests, the robot "died" because of a bad tuning of the salience computations: in situations characterized by a huge need of *Potential Energy*, when the robot approaches the *Potential Energy* resource, the dorsal loop increases, via STN, the output of all the channels of the ventral loop and stops the robot too far away from the desired resource. This shows the limitation of hand-tuning saliences and demonstrates the need for an adaptive learning mechanism capable of modifying the salience computations online.

5.2. *Coordination of the navigation strategies*

This experiment takes place in the environment ($6m \times 6m$) depicted in Fig. 2 (right), in which the robot may exhibit an opportunistic behaviour: when reaching the edge, it has to orientate towards an unexpected visible resource instead of following a planned path leading to a farther resource.

In a preliminary learning phase, the robot is allowed to build a topological map of this environment in which only the *Potential Energy* resource Ep1 is present, without metabolic constraints. In the test phase, a resource of the same nature, Ep2, is added to the environment and the initial *Potential Energy* level is set to 0.5. Following the planned path leading to Ep1, the robot begins to go to direction 0° but when it reaches the edge, it detects the new resource Ep2 and has to choose between two conflicting orientations: the planning vector leading to Ep1 (0°) or the object approach vector leading to Ep2 (270°).

Fifteen tests were run with various weights attributed to both planning and object approach vectors for salience computation. The results of Table 1 (right) show that, depending on which navigation strategy is favored, the robot is able to behave opportunistically thus choosing the closest resource (Ep2) or to ignore this unexpected resource and follow its initial plan (Ep1).

The indecision of the architecture, when both strategies have the same weight, is caused by the choice we made to merge orientations given by all the navigation strategies in the NAcc core and to select directions instead of selecting the strategies themselves. This alternative, suggested in a BG model [13], should be tested in a future implementation.

6. Discussion and Perspectives

We proposed an architecture of action selection integrating object approach and topological navigation strategies inspired by the structure and role of the basal ganglia in vertebrates. The advantage derived from the integration of the topological navigation strategy was demonstrated by an improved survival capability. The interest of using two different navigation strategies has been proven in a dedicated experiment.

From a neurobiological point of view, the model reaches the limits of current knowledge. Firstly, the coding of spatial information in the NAcc core being still undeciphered, a simplistic hypothesis of direction selection was made. The second experiment has suggested to try an alternative hypothesis of explicit navigation strategy selection. The link between these hypotheses and electrophysiological data remains to be considered. Secondly, the loop interconnection was achieved through the modelling of the trans-subthalamic pathway. However, the connections from one BG loop to another concern only a limited part of them, while, in our model, they concern the whole ventral loop output. Neurobiological data concerning the BG loops interconnections and their precise role is however currently too sparse to allow a more accurate model.

From the autonomous robotics point of view, this model proposes an integration of both action selection and navigation, taking into account the constraints of both issues. A preliminary experiment in a larger and more complex environment, including dangerous areas and in which various adaptive capacities were committed, showed that the robot can survive a very long time (up to 20 hours) [4]. The need to add learning abilities, as evoked in the first experiment, is taken into account in work currently in progress [9], which considers the role of dopamine as a reinforcement signal.

This model constitutes the first base of the control architecture of an "artificial rat", the *Psikharpax* robot. This integrative project aims at testing the coherency and the completeness of the current knowledge of the rat's nervous system and at building a robot which would be autonomous in the choice of its goals and actions.

References

[1] G. Chevalier and M. Deniau, Disinhibition as a basic process of striatal functions, *Trends Neurosci.*, **13**, 277 (1990).

[2] D. Filliat, Cartographie et estimation globale de la position pour un robot mobile autonome, *PhD. Thesis, Univ. Paris 6, France*, (2001).

[3] D. Filliat and J.-A. Meyer, Map-based navigation in mobile robots - I. A review of localization strategies, *J. Cog. Sys. Res.*, **4(4)**, 243 (2003).

[4] B. Girard, Intégration de la navigation et de la sélection de l'action dans une architecture de contrôle inspirée des ganglions de la base, *PhD. Thesis, Univ. Paris 6, France*, (2003).

[5] B. Girard, V. Cuzin, A. Guillot, K. N. Gurney and T. J. Prescott, Comparing a bio-inspired robot action selection mechanism with winner-takes-all, *SAB'02*, 75 (MIT Press, Cambridge, MA, 2002).

[6] A. Guillot and J.-A. Meyer, The Animat Contribution to Cognitive Systems Research, *J. Cog. Sys. Res.*, **2(2)**, 157 (2000).

[7] K. Gurney, T. J. Prescott and P. Redgrave, A computational model of action selection in the basal ganglia. I. A new functional anatomy, *Biol. Cyber.*, **84**, 401 (2001).

[8] K. Gurney, T. J. Prescott and P. Redgrave, A computational model of action selection in the basal ganglia. II. Analysis and simulation of behaviour, *Biol. Cyber.*, **84**, 411 (2001).

[9] M. Khamassi, B. Girard, A. Guillot and A. Berthoz, Mécanismes neuromimétiques d'apprentissage par renforcement dans l'architecture de contrôle du rat artificiel Psikharpax, poster presented at *plate-forme AFIA*, (2003).

[10] B. P. Kolomiets, J.-M. Deniau, P. Mailly, A. Menetrey, J. Glowinski and A.-M. Thierry, Segregation and convergence of information flow through the cortico-subthalamic pathways, *J. Neurosci.*, **21(15)**, 5764 (2001).

[11] P. D. Martin and T. Ono, Effects of reward anticipation, reward presentation, and spatial parameters on the firing of single neurons recorded in subiculum and nucleus accumbens of freely moving rats, *Behav. Brain Res.*, **116**, 23 (2000).

[12] N. Maurice, J.-M. Deniau, J. Glowinski and A.-M. Thierry, Relationships between the prefrontal cortex and the basal ganglia in the rat: physiology of the cortico-nigral circuits, *J. Neurosci.*, **19(11)**, 4674 (1999).

[13] O. Monchi, J. G. Taylor and A. Dagher, A neural model of working memory processes in normal subjects, Parkinson's disease and schizophrenia for fMRI design and prediction. *Neural Networks*, **13**, 953 (2000).

[14] J.-A. Meyer and D. Filliat, Map-based navigation in mobile robots - II. A review of map-learning and path-planning strategies, *J. Cog. Sys. Res.*, **4(4)**, 283 (2003).

[15] J. O'Keefe and L. Nadel, The Hippocampus as a Cognitive Map, *Oxford: Clarendon Press*, (1978).

[16] P. Pirjanian, Multiple Objective Action Selection Behavior Fusion using Voting, *PhD. Thesis, Aalborg Univ., Denmark*, (1998).

[17] J. K. Seamans, S. B. Floresco and A. G. Phillips, Functional differences between the prelimbic and anterior cingulate regions of the rat prefrontal cortex, *Behav. Neurosci.*, **109**, 1063 (1995).

[18] J. K. Seamans and A. G. Phillips, Selective memory impairments produced by transient lidocaine-induced lesions of the nucleus accumbens in rats, *Behav. Neurosci.*, **108**, 456 (1994).

STAGED LEARNING OF SACCADIC EYE MOVEMENTS WITH A ROBOT CAMERA HEAD

W. SCHENCK AND R. MÖLLER

Max Planck Institute for Psychological Research,
Cognitive Robotics Group,
Amalienstr. 33, 80799 München, Germany
E-mail: schenck@psy.mpg.de

In motor learning, two main problems arise: the missing teacher signal, and the necessity to explore high-dimensional sensorimotor spaces. Several solutions have been proposed, all of them limited in some respect. In the present work, an alternative learning mechanism is developed for the example of saccade control, implemented on a stereo vision robot camera head. This approach relies on two main principles: averaging over imperfect learning examples, and learning in multiple stages. In each stage, a saccade controller is trained with a set of imperfect learning examples. Afterwards, the output of this controller serves as starting point for the creation of a new training set with better quality. By the repetition of these steps, the controllers' performance can be incrementally improved without the need to search from scratch for the rare learning examples with very good quality.

1. Introduction

Internal models relate sensory signals from different modalities as well as sensory and motor signals. They are used to model functions of biological motor control systems on a computational level [5]. Internal models are devided in two broad categories: forward models and inverse models. Forward models predict the outcome of motor actions, while inverse models generate the necessary motor commands to attain sensory goal states. Controllers are inverse models with a constant goal state.

In this framework, motor learning is interpreted as training of adaptive internal models. In the following, we will focus on the training of controllers. From the current sensory state given as input, the controller produces a motor command as output (see Fig. 1). The training of controllers requires learning examples which pair the current sensory state with an appropriate motor command transforming the current sensory state into the constant goal state. These examples are difficult to collect. First, no teacher signal is available in motor space. While the deviation between the desired sensory state and the resulting sensory state after a movement is easily to obtain, in general the mapping of this error back to motor space is unknown. Thus, the sensory error cannot be exploited for learning. Alternatively, one could search systematically in sensorimotor space for good learning examples. However, in general this simple approach is computationally too expensive due to the high dimensionality of sensorimotor spaces.

To cope with both the missing teacher signal and the high dimensionality of sensorimotor spaces, several approaches to the learning of controllers (and inverse models) have been proposed: Feedback error learning [6], direct inverse modelling [4], distal supervised learning [4], and learning by interpolation [8]. Each approach has its own strengths and weaknesses which will be discussed later. In the present work, we suggest an alternative solution: *staged learning* in combination with *learning by averaging*.

We will illustrate the practical use of this approach in the domain of saccade control. Saccades are rapid eye movements (up to 900 deg/s in humans) for the fixation of interesting points in visual space. The fixated objects are projected onto the fovea, the region with highest cone density on the retina. Only when both eyes fixate the same object, the images of this object on both retinas can be fused to a single percept by the human brain. Since retinal delay is in the order of 50 msec (longer than the duration of most saccades), visual feedback is not fast enough to control saccades [3]. Therefore, the movement parameters for the eye muscles have to be precisely determined before a saccade starts. As research on infants shows [1], proper saccades towards peripheral targets have to be learnt during infancy. Thus, it is an important question how this learning works. In our study, we show the applicability of staged learning for saccade-like fixation movements of a robot camera head. This camera head is supposed to fixate target objects on a horizontal surface (a table).

2. Learning of Controllers

2.1. *Overview of Related Approaches*

In feedback error learning [6], a feedback controller is suggested for producing an approximate teacher signal in motor space. For saccade control, this approach works well [2, 3]. On the downside, one has to presume the a priori existence of a simple feedback control scheme which limits the general applicability of feedback error learning.

In direct inverse modelling [4], random motor commands are produced.

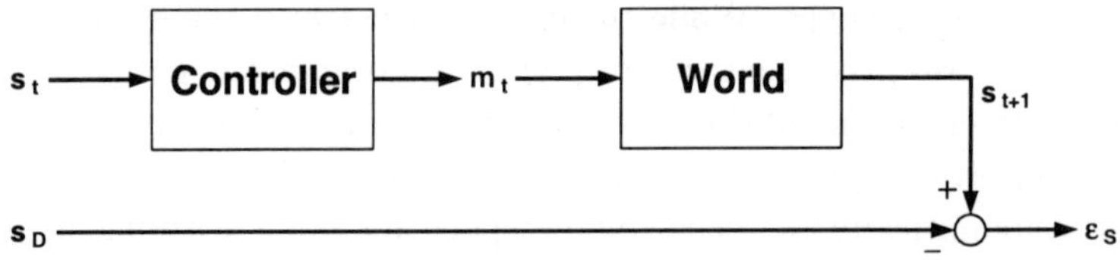

Figure 1. General controller input-output scheme (see text for details).

The sensory effect of each movement is incorporated in the respective learning example as desired sensory state while the movement command itself is used as motor output and the initial sensory state as sensory input of the learning example. This learning scheme suffers heavily from the high dimensionality of sensorimotor spaces. Controller learning is virtually impossible since one must search for movements which result exactly in the fixed sensory goal state of the control task.

Distal supervised learning [4] combines a controller with a forward model. First, the forward model has to be trained to predict sensory states resulting from the application of motor commands. Afterwards for the training of the controller, the teacher signal in motor space is generated by backpropagating the sensory error through the forward model. On the first glance, this seems to be an elegant and universal solution, but unfortunately the aquisition of a sufficiently precise forward model may suffer from the high dimensionality of sensorimotor spaces as well.

Learning by interpolation [8] is a more technical solution. The error in the resulting sensory state is transformed into a change in the initial sensory state before the movement. As a prerequisite, one needs to know how the mapping between both sensory states looks like. This is very unlikely for biological organisms.

2.2. *Learning by Averaging*

In our approach, learning by averaging is the principle underlying each stage in the learning process. Let s_t denote the sensory state at time step t, while s_D is the desired sensory state. In time step t, a controller C produces the motor command m_t. Due to this motor command, the world state changes, resulting in a new sensory state s_{t+1} (see Fig. 1). The difference between s_D and s_{t+1} is the sensory error ε_s. We define a function $Q_s(m)$ which quantifies the quality of motor command m applied in sensory state s: the lower ε_s, the higher $Q_s(m)$.

The basic idea of learning by averaging is to collect not only perfect

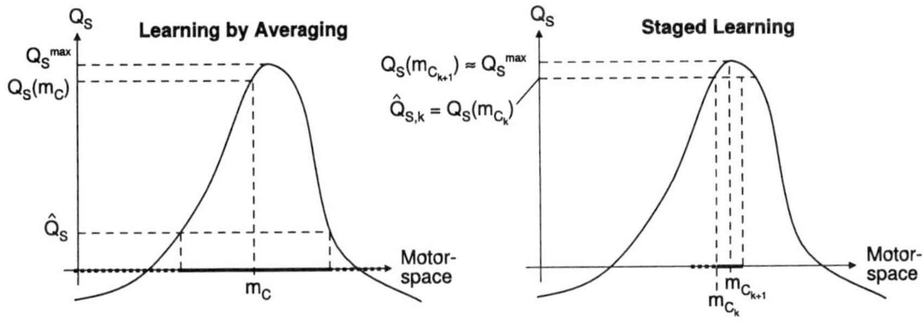

Figure 2. Learning by averaging (left) and staged learning (right). These graphs show the quality function $Q_s(m)$ over motor space for a fixed sensory controller input s. In motor space, both the bold and dashed area are sampled randomly, but only examples within the bold area on the X-axis are included in the training set.

learning examples, but to include every learning example in the training set whose quality exceeds a certain threshold $\widehat{Q}_s$ depending on the initial sensory state s. One obtains learning examples by sampling randomly in sensory and motor space. Whenever $Q_s(m)$ is larger than $\widehat{Q}_s$, the example (s, m) is included in the training set.

In the training, these imperfect learning examples are averaged. For this purpose, one needs a controller which adapts its output to the average of different motor commands corresponding to similar sensory inputs. A widely used architecture with such learning properties is the multi-layer perceptron. If the learning is successful, $Q_s(m_C)$ (m_C: Controller output) is close to maximum quality Q_s^{max} for all s (see Fig. 2, left).

The success of learning depends on the properties of $Q_s(m)$. Whenever $Q_s(m)$ is unimodal and symmetric for all s, success is guaranteed. The more $Q_s(m)$ deviates from this perfect form, the larger the expected difference between $Q_s(m_C)$ and Q_s^{max}.

2.3. *Staged Learning*

If $Q_s(m)$ is not unimodal and symmetric for all s, or if the number of learning examples is too small for proper averaging, the controller C that evolved from learning by averaging will not be precise enough. In this case, we suggest "Staged learning" as additional procedure. Let us consider the general case where we have a controller C_k trained in stage k. This controller produces output with a certain quality $Q_s(m_{C_k})$. For the generation of a new set of learning examples, sensory space is sampled randomly again. For each encountered sensory state s, a corresponding motor command is generated by adding noise to the controller output $m_{C_k}(s)$: $m = m_{C_k}(s) + \eta_k(m_{C_k}(s))$. Again, the example (s, m) is included in the training set only if the quality of the motor command $Q_s(m)$ exceeds a certain threshold $\widehat{Q}_{s,k}$. In staged learning, this threshold is computed by $\widehat{Q}_{s,k} = Q_s(m_{C_k})$ (see Fig. 2, right). It is often useful to approximate $\widehat{Q}_{s,k}$ because determining $Q_s(m_{C_k})$ requires the controller output m_{C_k} and the execution of the respective movement.

Based on controller C_k, a training set T_{k+1} is collected for stage $k+1$, and using T_{k+1}, a controller C_{k+1} is trained. By repeating this procedure, controller performance increases from stage to stage. When staged learning is combined with learning by averaging as suggested here, the controller of the first stage evolves as depicted in Section 2.2. Moreover, in controller training the collected learning examples are averaged.

Staged learning in combination with learning by averaging works fine, whenever the quality function $Q_s(m)$ is unimodal for all s. The more $Q_s(m)$ deviates from this ideal form, the more difficult staged learning will be (but still may work). With a well-chosen noise function $\eta_k(m_{C_k}(s))$, the great advantage of staged learning is the significant reduction of search effort for good learning examples as we will show for saccade control.

3. Saccade Controller

3.1. *Controller Design*

Our robot camera system consists of two cameras mounted on a single "pan-tilt unit" (PTU) with a horizontal (pan) and a vertical (tilt) rotation axis. The cameras are mounted around 85 cm above a table surface with a size of 80×80 cm. The distance between the optical axes of the cameras is approximately 20 cm. The task of the proposed saccade controller is to fixate interesting objects (colored wooden blocks) on a table surface in a way that the respective target object is centered in both camera images. In this study, 42 different wooden blocks were scattered over the table surface. The arrangement of blocks differed between controller training and testing.

Two motor parameters (pan and tilt) are controllable via hardware. A third parameter, the vergence angle between both cameras, determines the horizontal offset of two regions extracted from the camera images. Altogether, the saccade controller has three motor outputs: pan, tilt, and vergence. All these values are scaled to the range $[-1.0; +1.0]$. In pan and tilt direction, this range corresponds to just under 60 degrees.

Figure 3 shows the overall control scheme. The input of the controller

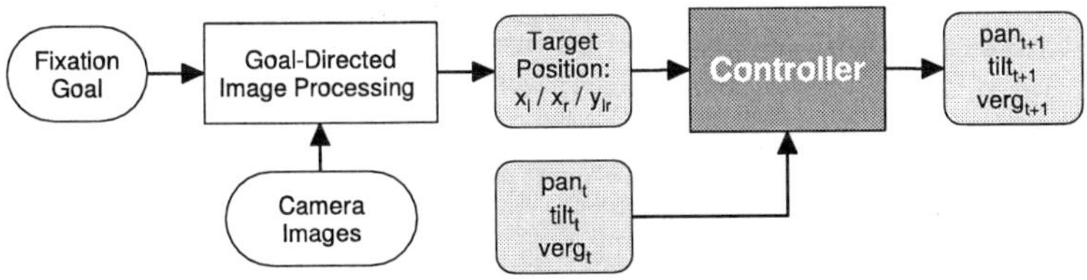

Figure 3. Overview of the saccade controller (see text for details).

is both kinesthetic and visual. The kinesthetic part consists of the current motor state in time step t (pan_t, $tilt_t$, and $verg_t$). The visual input is reduced to three values: x_l, x_r, and y_{lr}. These denote the position of the current target object in both the left and right "eye".[a] Thus, altogether there are six input values (also scaled to the range $[-1.0; +1.0]$).

Image processing takes the current camera images and the current fixation goal as input. The fixation goal determines which objects are most salient, depending on their size and color. In the first step of image processing, quadratic regions are cut out from the camera images and subsampled to 55×55 pixels (corresponding to a diagonal angle of view of 56 degrees). Afterwards, the most salient object on the table surface is identified as target object. The position of its center in both the left and right subsampled image is determined.

3.2. *Controller Training*

Controller training follows the procedure of staged learning as outlined in Section 2.3. Each learning example in the training set is generated in the following way: First, a random motor command is generated and executed (consisting of pan_t, $tilt_t$, and $verg_t$). Then, one of the wooden blocks on the table surface within the visual input of the controller system is randomly selected as target object. The image coordinates of its center are computed for both eyes (resulting in x_l, x_r, and y_{lr}). In this way, the complete controller input is determined. For output generation (pan_{t+1}, $tilt_{t+1}$, $verg_{t+1}$), one has to distinguish two cases. In stage one, no controller is available. The output is just a random variation of the initial motor position. From stage two on, first the output of the already existing controller from the last stage is determined (pan^*_{t+1}, $tilt^*_{t+1}$, $verg^*_{t+1}$). Then, the vector $[(pan^*_{t+1}, tilt^*_{t+1}) - (pan_t, tilt_t)]$ is scaled with a random value around

[a]Due to the geometry of the camera head, y_l and y_r are always equal; thus, they are represented by one value y_{lr}.

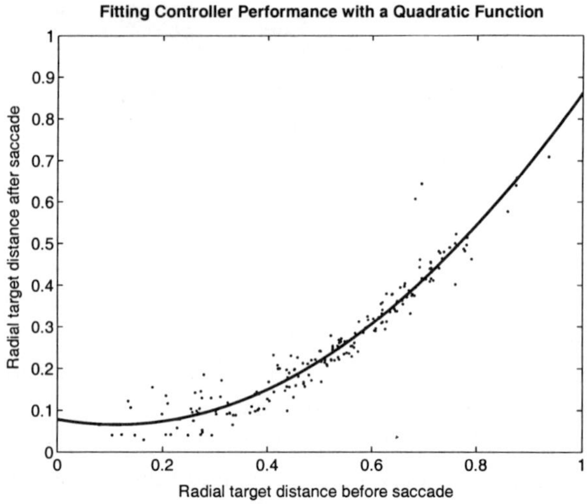

Figure 4. Points illustrate the performance of the first-stage controller resulting from learning by averaging. 200 random trials were carried out. The bold curve is a quadratic function fitting the functional relationship between the pre- and post-saccadic radial target distance (combined value for the left and right eye).

1.0. This scaled vector and random noise are added to $(pan^*_{t+1}, tilt^*_{t+1})$, resulting in the final endpoint $(pan_{t+1}, tilt_{t+1})$. $verg_{t+1}$ is determined by adding noise to $verg^*_{t+1}$. All types of noise are uniformly distributed, and noise amplitude is reduced from stage to stage. The exact definition of the noise function $\eta_k(m_{C_k}(s))$ does not seem to be critical for our specification of saccade control as long as the noise level decreases in each stage.

For the saccade controller, qualities and quality thresholds are always computed in pairs for the left and right side (Q^l and Q^r, in the following we will refer to both values as $Q^{l/r}$). Each random movement is only included as learning example in the training set if the quality of the movement is larger than a certain quality threshold on both sides. The quality $Q^{l/r}_{s_t}(m_t)$ of movement m_t in sensory state s_t is the negative of the radial target distance $r_{l/r}$ on the left/right side for the resulting sensory state s_{t+1}.[b] $r_{l/r}$ is computed as the Euclidean distance between the center of the subsampled input images and the target object coordinates in these images. $r_{l/r}$ is scaled to the range $[0.0; +1.0]$. For the first stage, the quality threshold $\widehat{Q}^{l/r}_{s_t}$ for sensory state s_t is the negative of $r_{l/r}$. Acceptable estimates of the correct motor command are obtained by averaging between both under-

[b]This quality function is unimodal over motor space for every sensory state.

and overshoot examples included in the resulting training set.

From stage two on, the quality threshold $\widehat{Q}^{l/r}_{s,k}$ is not exactly computed as $Q^{l/r}_s(m_{C_k})$. Instead, $\widehat{Q}^{l/r}_{s,k}$ is approximated (see Section 2.3) by fitting a quadratic function to the functional relationship between the radial target distance before a saccade, and after a saccade carried out by controller C_k (fitting is based on a combined radial target distance value for the left and right side, see Fig. 4). $\widehat{Q}^{l/r}_{s,k}$ is the negative of this fitted function. The data for the fitting procedure is collected by carrying out 200 random trials with controller C_k. This kind of approximation saves implementational and computational effort. The quadratic function was chosen for the fitting procedure since it captures the relationship between the pre- and post-saccadic radial target distance sufficiently well.

In each stage, 10,000 learning examples were collected for training. As controller network, a multi-layer perceptron with 40 hidden units was used since this network architecture is well suited for staged learning in combination with learning by averaging (see Section 2.2). As outlined in Section 3.1, the controller has six input units and three output units. Training was carried out with 2,000 epochs of resilient propagation [7].[c] To avoid outliers, in each stage five controller networks were trained. The controller with medium performance was taken as basis for the next stage.

4. Results

The left part of Table 1 shows the results for the sets of training examples. From stage to stage, the average radial target distance after each saccade in the training set decreases. In stage seven, it reaches 0.0. This means, that the respective training set consists exclusively of perfect learning examples, i.e. perfect saccades resulting in optimum fixation after only one move.

On the other hand, the average number of required random movements for the collection of one learning example increases from 4.2 to 19.1. Thus, the higher the performance level already reached, the more difficult it is to find even better learning examples. Altogether, in the overall learning history, for each of the perfect examples in the training set of stage seven, around 60 random movements had to be carried out.

To assess controller performance, first we tested fixation success (right part of Table 1). In each trial, a random motor starting position and a

[c]Resilient propagation needs significantly less epochs for training than standard back-propagation. Moreover, it has virtually no adjustable parameters and results in well-generalizing networks.

Table 1. Results over seven stages of learning, regarding both the sets of training examples and the trained saccade controllers. The results of the controllers with medium performance for each stage are reported here.

	Training set		Controller		
			Fixation		One-step
Stage	Random movements per pattern	Average $r_{l/r}$ after one saccade	Average number of needed saccades (SD)	Percent. of successful fixation trials	Average $r_{l/r}$ after one saccade
1	4.2	0.38 / 0.38	11.2 (5.05)	36 %	0.48 / 0.48
2	3.7	0.22 / 0.21	6.98 (3.99)	48 %	0.36 / 0.37
3	4.8	0.17 / 0.17	7.89 (5.11)	53 %	0.26 / 0.26
4	5.7	0.13 / 0.13	5.84 (4.04)	84 %	0.16 / 0.16
5	11.8	0.07 / 0.07	4.49 (3.88)	84 %	0.09 / 0.09
6	11.6	0.03 / 0.03	4.60 (3.56)	76 %	0.07 / 0.07
7	19.1	0.0 / 0.0	3.49 (3.01)	99 %	0.06 / 0.06

random target object were determined. Then, the controller carried out a series of saccades towards the target object until it was successfully fixated in the center pixel in both the left and right subsampled input image (this corresponds to a precision of 0.033% of the overall image area). The results in Table 1 are obtained from 200 fixation trials. The maximum number of saccades within one trial was restricted to 20. When this number was exceeded, the trial was counted as non-successful. The percentage of successful fixation trials increases from 36% in stage one to 99% in stage seven. The average number of needed saccades in one trial drops down from 11.2 to 3.49 (for successful trials). Moreover, we looked at the radial target distance after the first controller saccade. The average radial target distance decreases from 0.48 in stage one to 0.06 in stage seven.

5. Discussion

The results show that staged learning is successful in the domain of saccadic eye movements. Herein, the main result is the number of 60 random movements needed in the overall learning history for one of the perfect learning examples in the final training set. When searching from scratch for a single perfect example, around 60,000 movements[d] would be required

[d]The number of 60,000 random movements was determined by building up a training set as described in Section 3.2 for stage one. The quality threshold was set to a constant

instead. Thus, with staged learning, there is a huge gain in the speed of building up a very good training set. We expect that this gain will be the larger the higher the dimensionality of sensorimotor space.

Moreover, the missing teacher signal is replaced by a quality function over sensorimotor space. Here applies the restriction that ideally, this quality function is unimodal. In our ongoing work, we want to show that it is possible to define such quality functions for many different motor tasks, not only for saccade control. In this paper, we considered staged learning always in combination with learning by averaging. Actually, we think that staged learning will also work without averaging, e.g. in combination with recurrent neural networks. With such controller architectures, the quality function is not restricted to a unimodal form any longer.

If further research as outlined above is successful, staged learning will be a more general approach for the learning of controllers (and inverse models) than the procedures described in Section 2.1. Saccade control provides a promising start as shown in this work. Moreover, staged learning could be used to model human learning in many areas of motor control.

References

1. Aslin, R. N. (1987). Motor aspects of visual development in infancy. In P. Salapatek & L. Cohen (Eds.), *Handbook of infant perception* (pp. 43-113). New York: Academic Press.
2. Bruske, J., Hansen, M., Riehn, L., & Sommer, G. (1997). Biologically inspired calibration-free adaptive saccade control of a binocular camera head. *Biological Cybernetics,* ***77***, 433-446 .
3. Dean, P., Mayhew, J., & Langdon, P. (1994). Learning and maintaining saccadic accuracy: a model of brainstem-cerebellar interactions. *Journal of Cognitive Neuroscience,* ***6***, 117-138.
4. Jordan, M., & Rumelhart, D. (1993). Forward models: supervised learning with a distal teacher. *Cognitive Science,* ***16***, 307-354.
5. Kawato, M. (1991). Internal models for motor control and trajectory planning. *Current Opinion in Neurobiology,* ***9***, 718-727.
6. Kawato, M., Furukawa, K., & Suzuki, R. (1987). A hierarchical neural-network model for control and learning of voluntary movement. *Biological Cybernetics,* ***57***, 169-185.
7. Riedmiller, M., & Braun, H. (1993). A direct adaptive method for faster backpropagation learning: the RPROP algorithm. In *International Conference on Neural Networks* (pp. 586-591). San Francisco: IEEE.
8. Van der Smagt, P. (1995). *Visual robot arm guidance using neural networks.* Doctoral thesis: University of Amsterdam.

---

value (0.01) which allowed only perfect saccades to be included in the training set.

Developmental Processes

MODELLING ASYMMETRIC INFANT CATEGORIZATION WITH THE REPRESENTATIONAL ACUITY HYPOTHESIS

G. WESTERMANN* AND D. MARESCHAL

Centre for Brain and Cognitive Development
School of Psychology, Birkbeck College
Malet Street
London WC1E 7HX, UK

We present a neural network model that accounts for an observed asymmetry in the categorization of cats and dogs in 3–4 month old infants. The model establishes a link between infant behaviour and mechanisms of cortical processing. Based on developmental change in the cortex the model predicts behavioural change in infants between 3 and 10 months of age.

1. Introduction

What are the brain mechanisms underlying categorization? Computational models of categorization in infancy can shed light on this question in multiple ways. They have not only been able to account for different aspects of categorization behaviour [2–5] but they have also been successful in demonstrating how behavioural change can occur as the outcome of developmental changes in the cortex [11]. These models generally simulate infant behaviour in preferential looking tasks. In these tasks, infants are shown a sequence of pictures (or real objects) from one category (for example, pictures of cats) and the time the infants spend looking at these stimuli is measured. With continued exposure to these familiarization stimuli looking time normally decreases and the infants habituate to them. In a subsequent test phase, infants are presented with two pictures, one from the old category (e.g., another cat), and one from a new category (e.g., a dog). Infants prefer to look at novel things, and if they look, for example, longer at the novel dog but not the novel cat, it is assumed that they have formed a category of cats that includes the novel cat, but excludes the dog.

*current address: Department of Psychology, Oxford Brookes University, Oxford OX3 0BP, UK email: gwestermann@brookes.ac.uk

Infant behaviour in a variety of such categorization tasks has been successfully modelled in auto-encoder neural networks [2–5]. These are simple feed-forward backpropagation networks that learn to reproduce the input on the output side. Normally these models have a hidden layer that is smaller than the input and output layers, and the idea is that this bottleneck forces the model to develop a structured internal representation of the presented stimuli. The models are trained, like infants, on a set of familiarization stimuli and subsequently presented with test items. The network error (i.e., discrepancy between the stimulus representation and the network output) is seen as analogous to infant looking time (Fig. 1). This analogy is based on the idea that infant looking time is determined by the time necessary to build an accurate internal representation of an object by cycling through an encoding/comparison/adjustment cycle, and the more novel an object is, the longer this time will be [9]. In auto-encoder models, a more novel object will generate a higher network error which likewise would require more adaptation cycles than a familiar object.

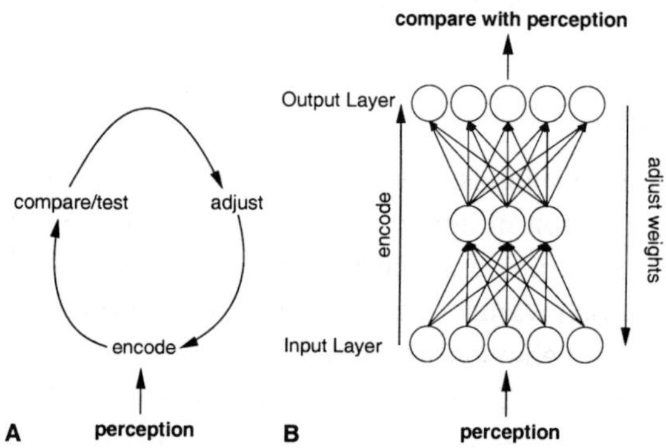

Figure 1. Learning by adjusting internal representations in infants (A) and in auto-encoder neural networks (B). (Adapted from [4].)

In the categorization of cats and dogs, a striking asymmetry has been found [8]: when 3–4 month old infants were familiarized on pictures of cats, they subsequently preferred a novel dog over a novel cat, indicating that they had formed a category of cats that included the novel cat and excluded the novel dog. However, when infants were familiarized on dogs, they showed no preference for either a novel cat or a novel dog, indicating that their "dog" category included novel cats as well.

This and related results have been modelled in a series of simulations with auto-encoder neural networks [2; 3; 5]. In these simulations, the cats and dogs were described in terms of ten key features (leg length, ear separation, ear length, nose length, nose width, head length, head width, eye separation, body length, body height). The models were trained on either 12 randomly selected cats or dogs, and were then tested on a novel cat and a novel dog. Mareschal *et al.* [5] found that their model displayed the same asymmetry as the 3–4 month old infants: after being trained on cats, a novel dog produced higher error than a novel cat. By contrast, after being trained on dogs, there was no significant difference between the errors for novel cats and dogs. Based on the model, Mareschal *et al.* [5] argued that the behaviour of the infants was determined by the perceptual characteristics of the stimuli: the distribution of the stimulus features showed less variance for cats than for dogs, and, importantly, the cat feature values were subsumed under the dog feature values. Mareschal *et al.* concluded that early infant categorization could be explained by perceptual bottom-up processes and was independent of higher-level category knowledge. Subsequent simulations and infant experiments have provided further support for this claim [2; 3; 7].

In this paper we present simulations of these experiments with a new model, the Representational Acuity Hypothesis Model [11]. These simulations present several improvements over the conventional auto-encoder model. In order to avoid catastrophic forgetting, the auto-encoder model had to be trained in batch mode, that is, all familiarization items were presented to the model before a weight update was made. However, this procedure does not correspond to the infant familiarization where stimuli were presented individually and it can be assumed that infants adapted to each presented stimulus when it was presented. Second, the auto-encoder modelled a static snapshot of the developmental stage of infants without accounting for transitions between behaviours. In fact the same model was argued to account for 10-month olds' behaviour in one study [4] and for the behaviour of 3–4 month olds in another [5]. Finally, in our model we want to make a more explicit link with cortical processing and show how developmental change can be explained by changes in cortical processing.

The Representational Acuity Model is based on the same principles as the previous auto-encoder, encoding familiarization stimuli and using the model error as an analogy to infant looking time in behavioural experiments. However, the internal structure of the model is different, leading to its ability to account for mechanisms of developmental change and to the identification of cortical mechanisms that underlie this change.

2. The Representational Acuity Hypothesis

The Representational Acuity Hypothesis (RAH) [11] was developed to account for the developmental change in categorization behaviour of infants from 4 to 10 months of age. This change has been argued to involve a shift from processing individual features at 4 months of age to processing relations between features at 10 months of age [12]. In these experiments, infants were familiarized on a set of animal pictures that varied along three feature dimensions (foot type, body type, tail type). Occurrence of individual features was correlated, that is, for example, a certain tail would always co-occur with a certain body. Infants were then tested on three animal pictures: one observed the feature correlation of the familiarization set, one had completely novel features, and one had previously used features, albeit in novel combinations that violated the correlational structure of the familiarization stimuli. In these experiments, 4-month olds did not look preferentially at this "switched" animal, whereas 10-month-olds did. This result was seen as evidence that 10-month olds had processed the correlations between features in the familiarization data and therefore considered the "switched" animal as novel, whereas 4-month olds had not [12].

We explained these results in our model based on the RAH [11]. We argued that objects are represented in higher cortical areas in terms of salient features. A neuron on this cortical "representational map" responds when the featural description of an object falls within its receptive field. In analogy with the development of visual acuity in the first months of life we argued that representational acuity develops based on a decrease of neuronal receptive field size during development. In young infants receptive fields are large and neurons respond to a variety of featural descriptions of objects. During development these receptive fields shrink and the neurons become tuned to more specific objects. Whereas rapid adaptations of receptive fields in response to particular stimuli have been reported (e.g., [10]), the RAH is concerned with the more gradual decrease in receptive field size that occurs over a timescale of months during early development (e.g., [1]).

We modelled the RAH with an auto-encoder neural network that had a hidden layer of units with Gaussian activation functions (in effect, an auto-encoder radial basis function network). The hidden layer modelled the representational cortical map. The stimuli used in the infant experiments [12] were encoded in terms of their features and presented to the model which was trained to reproduce these inputs on the output side. To model the development of representational acuity, the model was trained with different widths of the hidden Gaussian activation functions. Younger infants were modelled with large receptive fields, and older infants with small receptive fields. Like in the conventional auto-encoder [4], network error for test items was considered analogous to infant looking time for these test items.

Training the model with different receptive field sizes accounted for the behavioural shift that was observed with infants of different ages [11]. With large receptive fields, the model displayed the same error for the previously seen and the "switched" test stimulus, corresponding to the behaviour of 4-month olds, whereas with small receptive fields, error for the switched item increased to be as high as that for the novel test item, corresponding to the behaviour of 10-month olds.

3. Modelling Asymmetry in Categorization

In this paper we extend our simulations with the RAH model as a biologically more plausible alternative to the conventional auto-encoder model and show how the categorization behaviour of 3–4 month old infants can be explained with recourse to the development of cortical receptive fields. Further, by varying the size of receptive fields in the model we can make prediction about the categorization behaviour of older infants.

3.1. *Cat–Dog Asymmetry*

The first study addressed the original results of cat–dog asymmetry [8] that have previously been modelled using the auto-encoder model [5]. We used the same data as this earlier model. The RAH auto-encoder had 10 linear input and 10 linear output units. 400 Gaussian hidden units were randomly placed in the 10-dimensional input space. To simulate young infants with large cortical receptive fields a receptive field size of 3.8 ($1.2\sqrt{d}$ with $d = 10$ the dimensionality of the input space) was used. For cat-familiarization trials, the model was trained by presenting 12 randomly selected cats in pairs of two for 1000 iterations (weight updates) before proceeding to the next pair. For dog-familiarization trials, the model was likewise trained on 12 randomly selected dogs in pairs of two. Weights from the hidden to the output units were adapted at each iteration with the perceptron learning rule. Receptive field positions (that is, input-hidden weights) were not adapted. After training, the model was tested on a novel cat and a novel dog. Each simulation was run 50 times with different random weights and hidden unit positions.

Fig. 2A shows the result of this simulation. When the model was trained on cats, test error for a novel dog was significantly higher than for a novel cat. By contrast, when trained on dogs, test error for novel cats and novel dogs was not significantly different. This result models the experimental data from 3–4 month old infants [8].

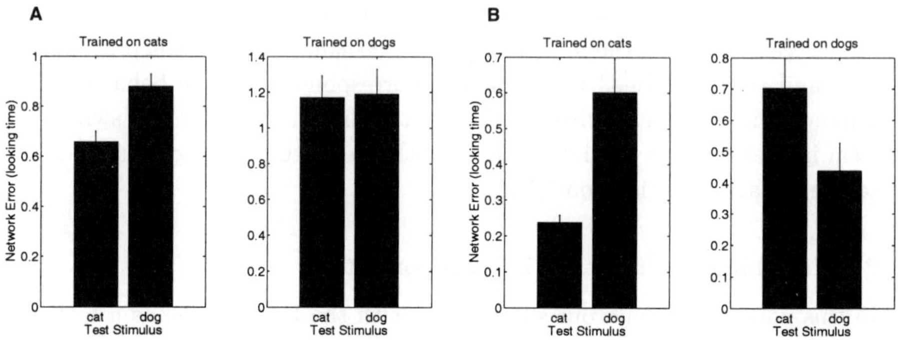

Figure 2. Simulation of the asymmetry between cat and dog categorization with large (A) and small (B) receptive fields. Standard error bars are also plotted. Results are averaged over 50 simulations each.

Why does the model display the asymmetry in categorization? As shown by Mareschal *et al.* [5], in most of the relevant feature dimensions the features for cats show less variability than those for dogs and are subsumed under the dog features. In effect, all cats look rather similar to each other, whereas dogs can look quite different from each other. On the representational feature map this means that the neurons that are activated by representations of cats lie within the area of neurons that are activated by dogs (Fig. 3A). As a consequence, when the model is trained on dogs the cat receptive fields are also (partially) tuned to the presented stimuli because the large receptive fields for cats and dogs overlap in this region (Fig. 3B). Subsequent presentation of a cat therefore does not lead to a higher error than presentation of a dog. However, when the model is trained on cats, only a small subset of the dog receptive fields are tuned (Fig. 3C), and subsequent presentation of a novel dog leads on average to higher error than a novel cat.

The asymmetry in the model exists because the receptive fields of neurons responding to cats and dogs overlap. This overlap is due to their large size, simulating an early stage of infant development. We have argued that

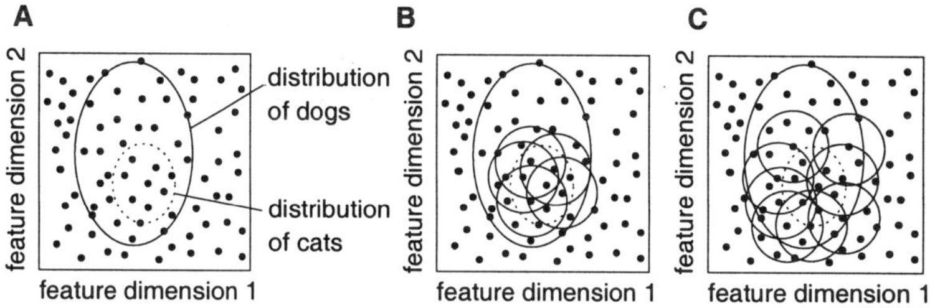

Figure 3. A. An illustrative example of the distribution of cats and dogs on the representational map. B. Cat receptive fields that are tuned by responding to dogs. C. Dog receptive fields that are tuned by responding to cats. For illustrative purposes here the representational map is assumed to represent only two feature dimensions (instead of ten).

the decrease in receptive field size in the cortex is a mechanism that leads to behavioural change in infant categorization [11]. In order to model the categorization of cats and dogs by older (10-month old) infants, we repeated the simulation exactly as described above, with the only difference that the size of the hidden unit receptive fields was reduced to 0.63 ($0.2\sqrt{d}$). The result of this simulation is shown in Fig. 2B. The categorization asymmetry has disappeared: when trained on cats, as before, error is higher for a novel dog than for a novel cat. However, now there is also a higher error for novel cats when the model has been trained on dogs. The model thus predicts that the category asymmetry in 3–4 month old infants disappears with increasing age, due to the decrease in receptive field size on a representational cortical map. This result has not been explicitly tested in preferential looking experiments, however, object examination studies have found that infants form exclusive cat and dog categories by 11 months [6].

3.2. *Reversing the Asymmetry*

To verify the claim that early infant categorization is based on perceptual bottom-up processes and depends on the feature distribution in familiarization and test stimuli, French *et al.* [3] showed that the cat-dog asymmetry could be reversed by carefully selecting and slightly distorting pictures of cats and dogs so that now, cat features had a broader variance than dog features and subsumed the dog features. They predicted that with this new stimulus set, infants should show a reversed asymmetry, preferring a novel cat when familiarized on dogs, and showing no preference when familiarized on cats. Their prediction was confirmed in experiments with 3–4 month olds as well as in simulations with the auto-encoder model.

The RAH model was also able to account for these results. Fig. 4A shows the network error when trained on the new data in the same way as above, with large receptive fields. Now, when trained on dogs, network error is higher for a novel cat than for a novel dog, but when trained on cats, there is no significant difference between the errors.

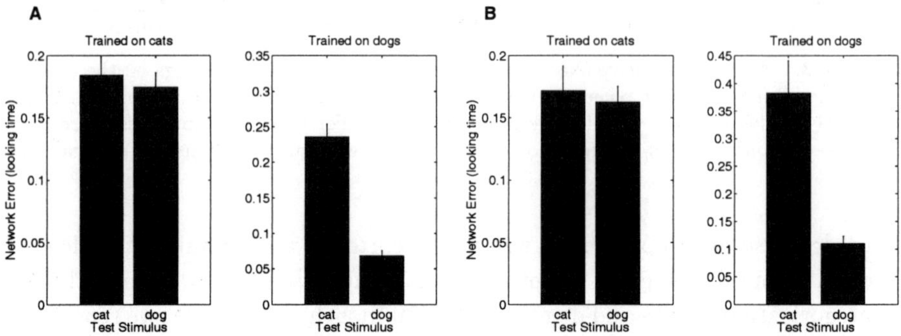

Figure 4. Modelling the reversal of asymmetry with large (A) and small (B) receptive fields. Standard error bars are also plotted. Results are averaged over 50 simulations each.

Modelling older infants in the asymmetry reversal condition by using smaller hidden unit receptive fields (Fig. 4B) yielded the surprising result that here the asymmetry did not disappear. The reason for this result is that familiarization on cats did not decrease the error for novel cats in testing. This might be due to the fact that the cats in the asymmetry-reversal condition are more varied than the dogs in the asymmetry condition. Infant studies with 10-month olds will have to test this prediction.

3.3. *Eliminating the Asymmetry*

In a final study [2], French *et al.* eliminated the asymmetry in young infants by again choosing carefully and distorting cat and dog pictures in a way that their feature distributions did no longer overlap. Infant experiments and simulations with the auto-encoder model showed that with this data set, the asymmetry did indeed disappear, providing further strong evidence for the bottom-up feature-based view of infant categorization.

We modelled this result in the RAH model by training it as before, but this time with the non-overlapping stimuli. As before, to model younger infants we used a large receptive field size. The results of this simulation are shown in Fig. 5. In this condition, even with large receptive fields, the model shows a symmetric categorization, with novel dogs preferred over (resulting in higher error than) novel cats when trained on cats, and novel cats preferred over novel dogs when trained on dogs.

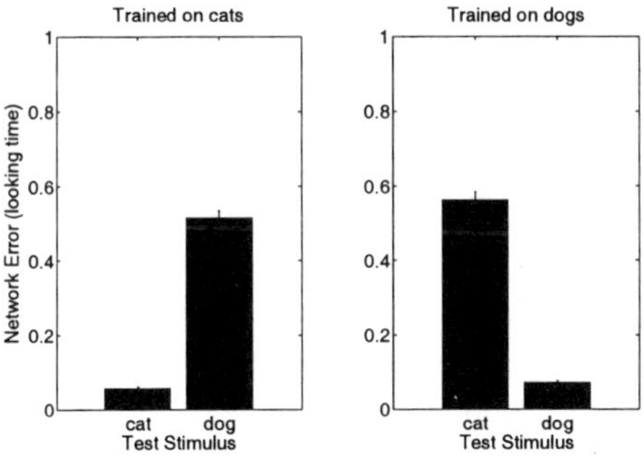

Figure 5. Modelling the suppression of asymmetry in younger infants with large receptive fields. Standard error bars are also plotted. Results are averaged over 50 simulations.

4. Discussion

The simulations reported in this paper provide further support for the Representational Acuity Hypothesis [11]. The RAH was implemented in a simple neural network model that, in varying the size of the hidden unit receptive fields, makes direct reference to developmental changes in the brain to produce behavioural change in infant categorization. The model can account for observed behaviours in the categorization of cats and dogs in 3–4 month old infants, and it makes predictions about the behaviour of 10-month olds. Together with the simulations of behavioural change leading from the processing of individual features to correlations between features [11], the RAH provides a parsimonious account of the link between brain and cognitive development in early infant categorization.

Acknowledgments

This work was supported by European Commission RTN grant HPRN-CT-2000-00065 and by ESRC (UK) grant R000239112.

References

1. B. O. Braastad and P. Heggelund. Development of spatial receptive field organization and orientation selectivity in kitten striate cortex. *Journal*

of Neurophysiology, 53:1158–1178 (1985)

2. R. M. French, D. Mareschal, M. Mermillod, and P. Quinn. The role of bottom-up processing in perceptual categorization by 3- to 4-month old infants: Simulations and data. (under review).
3. R. M. French, M. Mermillod, P. Quinn, and D. Mareschal. Reversing category exclusivities in infant perceptual categorization: Simulations and data. In *Proceedings of the 23nd Annual Conference of the Cognitive Science Society*, pages 307–312, Hillsdale, NJ: Erlbaum (2001).
4. D. Mareschal and R. M. French. Mechanisms of categorization in infancy. *Infancy*, 1:59–76 (2000).
5. D. Mareschal, R. M. French, and P. C. Quinn. A connectionist account of asymmetric category learning in early infancy. *Developmental Psychology*, 36:635–645 (2000).
6. D. Mareschal, D. Powell, and A. Volein. Basic-level category discrimination by 7- and 9-month-olds in an object examination task. *Journal of Experimental Child Psychology*, 86:87–107 (2003).
7. D. Mareschal, P. C. Quinn, and R. M. French. Asymmetric interference in 3- to 4-month-olds' sequential category learning. *Cognitive Science*, 26:377–389 (2002).
8. P. C. Quinn, P. D. Eimas, and S. L. Rosenkrantz. Evidence for representations of perceptually similar natural categories by 3- and 4-month-old infants. *Perception*, 22:463–475 (1993).
9. E. N. Sokolov. *Perception and the conditioned reflex*. Erlbaum (1963).
10. K. Suder, K. Funke, Y. Zhao, N. Kerscher, T. Wennekers, and F. Wörgötter. Spatial dynamics of receptive fields in cat primary visual cortex related to the temporal structure of thalamocortical feedforward activity. *Experimental Brain Research*, 144:430–444 (2002).
11. G. Westermann and D. Mareschal. From parts to wholes: Mechanisms of development in infant visual object processing. *Infancy*, 5 (2004).
12. B. A. Younger and L. B. Cohen. Developmental change in infants' perception of correlations among attributes. *Child Development*, 57:803–815 (1986).

A COMPUTATIONAL MODEL OF THE EMERGENCE OF GAZE FOLLOWING

ERIC CARLSON AND JOCHEN TRIESCH

Cognitive Science Department, UC San Diego,
9500 Gilman Drive, La Jolla, CA 92093-0515, USA
E-mail: {eccarlson,triesch}@ucsd.edu

We present a computational model of the emergence of gaze following in infant caregiver interactions. Using the model we explore the plausibility of the hypothesis that gaze following is a skill that infants acquire because they learn that monitoring their caregiver's direction of gaze allows them to predict where interesting objects or events in their environment are. In particular, we demonstrate that a specific *basic set* of mechanisms is sufficient for gaze following to emerge and we show how plausible alterations of model parameters motivated by findings on developmental disorders lead to impairments in the learning of gaze following.

1. Introduction

The capacity for shared attention is a cornerstone of human social intelligence. It is an early foundational skill of vital importance for learning in social contexts. By the end of their first year of life, normally developing infants show robust gaze following[4] as well as a variety of other shared attention skills, such as pointing and requesting behaviors. How do these skills develop? Why is their emergence disrupted in certain developmental disorders? Our particular focus is the emergence of gaze following behavior. We propose a computational model for the emergence of gaze following based on reinforcement learning. The model demonstrates that gaze following behavior can be robustly learned with a generic *basic set* of mechanisms[6]. Our learning-based view is in stark contrast to earlier nativist/modularist accounts of the emergence of shared attention skills[1,11] and offers very different explanations for failures of the emergence of gaze following in developmental disorders such as autism and Williams syndrome.

The work described in here is part of the MESA project at UC San Diego (**M**odeling the **E**mergence of **S**hared **A**ttention, `http://mesa.ucsd.edu`), a larger effort to understand the emergence of shared attention in normal and abnormal development through closely integrating observational studies with computational modeling approaches[6].

2. The Basic Set Hypothesis

We start from the premise that earlier nativist/modularist theories of the emergence of shared attention skills[1] are of little explanatory power. The problems of such theories have been eloquently pointed out by many researchers[14,5] and will not be discussed here. We believe that better theories will be based on a careful analysis of the developing infant's *interactions* with her environment during development. For the case of of gaze following, Moore has suggested that gaze following may emerge because infants learn that the caregiver's direction of gaze is predictive of where interesting things are in the environment[10]. This idea is at the core of our model. Our goal is to formalize this hypothesis into a concrete mathematical model and to test its plausibility and implications using computer simulations.

Previously, we hypothesized that gaze following may emerge from an interplay of a *basic set* of mechanisms[6] which may contain the following:

- a set of perceptual and motivational biases and habituation mechanisms driving the infant to look at and shift attention between "interesting" visual stimuli
- a generic hedonistic learning mechanism that learns behavioral strategies to satisfy the above preferences
- a structured environment providing correlations between where caregivers look and where interesting stimuli are.

We will briefly discuss each component in turn. Evidence that these mechanisms are indeed in place at the time that gaze following starts to emerge is reviewed in Fasel *et al.*, 2002.

Perceptual Preferences. Visual preferences of infants have been well studied in various kinds of preferential looking paradigms. A key component of our model is a strong infant social orientation. Normally developing infants enjoy looking at faces in general and their caregivers in particular. A further assumption that we are making is that infants' face processing skills allow them to distinguish different caregiver head poses by the time that gaze following emerges.

Habituation. We propose that the chief reason for the infant to shift gaze from one target to another is that due to habituation it is more "rewarding" for the infant to shift gaze to a different object after looking at one object for some time. Behavioral results suggest that between 8 and 12 weeks, infants begin to shift attention away from a central stimulus when a peripheral stimulus is introduced[2]. At six month infants begin breaking mutual gaze to look at distal objects.

Reward Driven Learning. We hypothesize that a hedonistic learning mechanism, *i.e.* a learning method seeking to maximize the pleasure or reward of the infant, is important for gaze following to emerge. In this sense, gaze following is *useful* for the infant since it allows the infant to maximize the amount of rewarding visual stimuli that she will see.

Structured Environment. A critical assumption of our model is that the infant's environment shows strong correlations between where the caregiver is looking and where interesting things are. In general, people tend to look at other people, or their own hands, while manipulating objects[8]. This tendency should introduce exactly such a correlation. In a parallel study, we are currently collecting observational data from naturalistic mother infant interactions to investigate this point more carefully.

How can these elements of the basic set work together to allow for gaze following to emerge? The infant's social predispositions will lead to frequent face-to-face interactions with the caregiver. Due to habituation the infant will more or less frequently look away from the caregiver's face. If the infant happens to look at the location looked at by the caregiver, she is somewhat more likely to see something interesting than if she shifts gaze to a different location. This allows the infant to establish an association between the caregiver's head pose and the "correct" gaze shift likely resulting in a rewarding sight.

3. Computational Model Based on Reinforcement Learning

In the following we present a simple computational model to demonstrate how the mechanisms of the basic set lead to the emergence of gaze following. To this end, we formalize our basic set of mechanisms in the framework of a biologically plausible reinforcement learning method — temporal difference learning. The goal of this exercise is to illustrate that the small basic set of plausible mechanisms is indeed *sufficient* for the emergence of gaze following to occur. We do *not* suggest, however, that these mechanisms are necessarily the *only* ones that are important for a comprehensive theory of the emergence of gaze following.

Our infant model is that of a pleasure-driven observer. There are many ways of formalizing this idea but a particularly appropriate formal framework is reinforcement learning[13]. In reinforcement learning, an agent interacts with its environment, occasionally receiving rewards, which it tries to maximize. In particular, we are interested in temporal difference learning approaches (TD learning), which have recently been associated with some aspects of the function of the basal ganglia[12].

In reinforcement learning models, at any time t the world (agent and environment) is in a particular *state* s_t. Time progresses in discrete steps ($t = 0, 1, 2, \ldots$) corresponding to roughly 250 ms. Upon observing the current state s_t, the agent decides to take an *action* a_t and potentially receives a *reward* r_t as a consequence. The potentially probabilistic mapping from state to action is the agent's *policy*, which is adapted during learning. The goal of the agent is to maximize its *future discounted reward*

$$R_t = \sum_{k=0}^{\infty} \gamma^k r_{t+k+1} \,, \tag{1}$$

where r_{t+k+1} is the reward received at time $t+k+1$, and $0 \leq \gamma \leq 1$ is a so-called *discount factor*. The discount factor determines how heavily the agent values the prospect of future rewards over immediate rewardsand lets them affect its behavior.

For simplicity, we assume that infant and caregiver are facing each other and remain in the same position. The caregiver can look either at the infant or at a set of $N = 10$ discrete regions in space. The infant can look at the caregiver or at any of the same regions. This shifting of gaze is the only way the infant model interacts with the environment. Thus, we predict that locomotion and physical interaction with the environment — although potentially very important — may not be strictly *necessary* for the emergence of gaze following.

Caregiver and Environment Model. In the environment there is always one interesting target present at any time in one of the N regions. There is a certain probability p_{valid} that the caregiver is looking at the location of that target. If the caregiver is not looking at the target, the caregiver's new direction of gaze is drawn from a uniform distribution over all of the other N locations (one for the infant plus N-1 regions of space not containing the target).

After some minimal fixation time at one location $T_{\text{fix}} = 4$ (1.0 sec.) the interesting target is relocated with some probability $p_{\text{shift}} = 0.5$ at each time step. The target is moved to a different location and a new direction of the caregiver's gaze is determined as described above. The parameters above were chosen *ad hoc* but will ultimately be set to match the findings from our observational studies on infant caregiver interactions.

Infant Model. The infant model receives rewards for looking at interesting things. The amount of reward received depends on the contents of the infants gaze and how habituated the infant is to those contents. There are four possible things for the infant to see, a frontal view of the caregiver (in case the caregiver is also looking at the infant), a profile view of the caregiver (in case the caregiver is not looking at the infant), an interesting object, or no object. Associated with these are base rewards $R_{\text{frontal}}, R_{\text{profile}}, R_{\text{object}}, R_{\text{nothing}}$. The actual reward received by the infant, or 'instantaneous reward', is the base reward mediated by habituation. As the infant looks at a location, the target, or the caregiver, the infant habituates to it. Similarly, the infant dishabituates to any objects it is not currently looking at.

when the infant model is not looking at something the infant dishabituates to it. For each location and object in the environment, including the caregiver, the infant has a habituation value, represented as a percentage of the potential reward. As the infant continues to fixate on an object its habituation value decreases. The habituation level to the current object of fixation at any time step is given by $h_{\text{fix}}(t) = h_{\text{fix}}(0)e^{-t}$, where $h_{\text{fix}}(0)$ is the habituation level at the beginning of the current fixation, and t is the time since the start of the fixation. Thus, the instantaneous reward actually received by the infant at time t is $r_{\text{inst}}(t) = R_{\text{fix}} h_{\text{fix}}(t)$, where $R_{\text{fix}} \in \{R_{\text{frontal}}, R_{\text{profile}}, R_{\text{object}}, R_{\text{nothing}}\}$ is the base reward for the object being fixated.

Likewise, when the infant is not looking at an object it dishabituates according to $h_{\text{location}}(t) = 1 - h_{\text{fix}}(0)e^{-t}$, where t is the time since last looking at the location and h_{object} is the level of habituation to that location.

Based on the work of Findlay & Walker[7], who propose that the decisions of when to shift gaze and where to shift gaze are made in somewhat separate neural pathways, the infant model consists of two separate agents. The state space of the 'when-agent', which decides whether to continue to fixate on the same location or shift gaze, has two dimensions. The first dimension represents the time the infant has been fixating at the same location discretized as the number of time steps (0, 1, 2, ..., 8, 9 or above). The second dimension is the instantaneous reward received by the infant. This is the total reward the infant receives on that time step taking habituation into account discretized evenly into ten discrete regions between the maximum and minimum possible rewards.

If the when-agent makes the decision to shift gaze, the 'where-agent' determines the target of the gaze shift. The state space of this agent has only a single dimension: the caregiver's direction of gaze, which can be any of the N regions of space, the infant, or unknown. Unless the infant is looking at the caregiver, the caregiver's direction of gaze will be unknown to the infant.

The agents use a Temporal Difference algorithm to estimate so-called *state-action values* $Q_t(s,a)$. These are estimates of the future discounted reward the agent will receive when choosing action a in state s and following the current policy thereafter. Estimation of the $Q_t(s,a)$ is done with the Sarsa algorithm[13]. On taking an action and receiving a reward, the *temporal difference* — the difference between the estimated future discounted reward of the next state plus any immediate reward received and the current estimated value of that state-action pair — is computed as

$$\delta_t = r_t + \gamma Q_t(s_{t+1}, a_{t+1}) - Q_t(s_t, a_t) \,, \tag{2}$$

where $Q(s_t, a_t)$ is the state-action value assigned to the state-action pair (s,a) at time t. In the Sarsa algorithm this temporal difference is used to adjust the state-action value by

$$Q_{t+1}(s,a) = Q_t(s,a) + \alpha\delta_t \,, \tag{3}$$

where α is a learning rate parameter.

We still need to specify how the infant model chooses its actions based on the estimated action values, i.e., chooses where to look next. It turns out that always choosing the action with the highest Q-value for the current state is typically not a good idea because it may lead to the agent "getting stuck" with a suboptimal policy early on and never discovering a better solution. It is typically better for the agent to sometimes chose exploratory actions that appear suboptimal given the agent's current knowledge, in order to get a better estimate of their true value. This is know as the *exploration vs. exploitation dilemma* in reinforcement learning. Our infant model chooses actions in a probabilistic manner using a *softmax decision rule*. Action a is chosen with probability:

$$p_t(a \mid s) = \frac{e^{\tilde{Q}_t(s,a)/\tau}}{\sum_{a'=1}^{N} e^{\tilde{Q}_t(s,a')/\tau}} \tag{4}$$

where $\tilde{Q}_t(s,a) = Q_t(s,a)/\max_{a'} |Q_t(s,a')|$ and τ is a so-called *temperature* parameter. The temperature determines how random the model's actions will be, *i.e.* how much it values exploration over exploitation. The lower τ, the more likely it is for the model to chose the action with the highest Q-value. In the extreme case of $\tau \to 0$ the model will follow a *greedy* policy, i.e. it will always chose the action with the highest Q-value. For $\tau \to \infty$, the model will chose all actions with equal probability. Hence, varying τ amounts to varying how the agent trades off exploration vs. exploitation.

4. Experiments

In this section we describe a first analysis of the model and the influence of some of the model's parameters on its learning behavior. In all of the following experiments we will set the learning rate to $\alpha = 0.0025$, the discount factor to $\gamma = 0.8$, and the temperature parameter of the soft-max action selection to $\tau = 0.095$. Our standard set of reward values are $R_{\text{frontal}} = R_{\text{profile}} = R_{\text{object}} = 1$ and $R_{\text{nothing}} = 0$. The probability of the caregiver looking to the location of the target when making a shift is set to $p_{\text{valid}} = 0.75$.

To quantify the emergence of gaze following in the model and its dependence on model parameters we use the following approach. At specific points during the learning process we temporarily "freeze" the model and evaluate its behavior for 1000 time steps (which corresponds to slightly more than 4 minutes), after which the learning process resumes. The model behavior at these stages of the learning process is analyzed by observing the infant model interacting with the environment and computing two statistics. The *caregiver index* CGI is defined as the frequency of gaze shifts towards the caregiver:

$$\text{CGI} = \frac{\#\text{ gaze shifts to caregiver}}{\#\text{ gaze shifts}} . \tag{5}$$

The *gaze following index* GFI is the frequency of gaze shifts that lead from the location of the caregiver to where the caregiver is looking:

$$\text{GFI} = \frac{\#\text{gaze shifts from caregiver to location looked at}}{\#\text{ gaze shifts}} . \tag{6}$$

An example run of the system is shown in Fig. 1 left. The model quickly learns to look at the caregiver. Gaze following behavior emerges somewhat later.

4.1. *Modeling Autism and Williams Syndrome*

While traditional nativist/modularist accounts typically propose broken or missing modules as the origin of developmental disorders[1], our learning

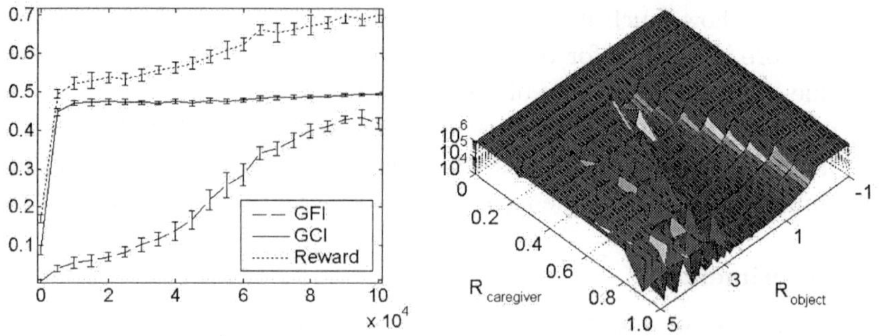

Figure 1. Left: Emergence of gaze following. The solid curve plots the CGI, the dashed curve plots the GFI, and the dotted curve plots mean reward per time step, as functions of learning iterations. Right: Learning performance as a function of caregiver- and object reward. The z-axis is the number of learning iterations necessary to achieve a GFI $\geq$ 0.3.

based account prompts us to look for potential differences in the components of the basic set that may lead to different developmental trajectories. For autism spectrum disorders and Williams syndrome, a particular interesting candidate is the reward structure of the model, since it is well established that autistics tend to avoid looking at faces while Williams syndrome patients are reported to show an abnormally high preference for looking at faces. Due to space limitations, this will be the only variation reported here.

We systematically varied the reward structure, i.e. parameters R_{frontal}, R_{profile} and R_{object} over a range of values with $R_{\mathrm{profile}} = R_{\mathrm{frontal}}$. Figure 1 (right) summarizes the results. For $R_{\mathrm{object}} \leq 0$ no gaze following behavior emerges. This makes intuitive sense because if the objects/events that the caregiver tends to look at are not rewarding for the infant, there is no advantage in gaze following behavior. More interestingly, if R_{frontal} and R_{profile} are negative, modeling aversion to faces as seen in autism, gaze following behavior does not develop normally. Interestingly, an analysis of the model shows that it will nevertheless slowly learn how to follow gaze but only rarely executes this behavior since it is unrewarding to do so. This matches a puzzling finding by Leekam *et al.*[9] demonstrating that autistics are often quite capable of following gaze if explicitly told to do so, although they may not exhibit the behavior normally.

To model Williams syndrome, we altered the relative strength of reward for looking at the caregiver vs. looking at the object. If the reward for looking at the caregiver is much higher than the reward for looking at the target, we found that gaze following is substantially delayed, as observed in Williams syndrome.

5. Discussion

We have presented a computational model of the emergence of gaze following. The underlying idea of this model is that infants learn to follow gaze because their caregivers tend to look at interesting targets that the infant will find rewarding to look at. If, indeed, the caregiver does this, the formal model demonstrates that a generic basic set of mechanisms is indeed *sufficient* for the emergence of gaze following in infant caregiver interactions. This is not to say that there are no other mechanisms playing an important role in the emergence of gaze following, it merely says that such mechanisms should not be strictly *necessary* for the emergence of gaze following.

A systematic analysis of the model's learning process and its dependence on system parameters showed that gaze following behavior emerges robustly for wide ranges of parameter values. But also, we found that changes to the reward structure of the model could lead to deficits in the learning of gaze following reminiscent of autism and Williams syndrome. We consider the model's ability to produce disruptions of the emergence of gaze following as seen in two radically different developmental disorders as quite attractive. In essence, we feel that our learning based account may provide a more parsimonious explanation of the emergence of gaze following than previous nativist/modularist approaches such as Baron-Cohen's.

Of course, our model must be seen as only a first step towards a full computational account of the emergence of gaze following. In many respects, the model is still overly simplistic. Examples are the restriction to a small set of discrete spatial regions and the absence of peripheral vision, the stereotypic, non-interactive behavior of the caregiver model, and the restriction of having only one interesting target present at any time, just to name a few. Nevertheless, we feel that our simple model is a useful tool in theorizing about the emergence of gaze following. In some respect, the utmost simplicity of the model can even be seen as a strength, since it brings the computational essence of the underlying mechanisms into focus. However, in order to account for additional experimental findings, such as the stage-wise development observed by Butterworth and Jarrett[3], some of the above-mentioned short-comings will be addressed in future work.

6. Acknowledgments

This work is supported by the UC Davis M.I.N.D. Institute and the National Alliance for Autism Research. We thank Gedeon Deák, Javier Movellan, Hector Jasso, and Leigh Sepeta for fruitful discussions.

References

1. Simon Baron-Cohen. *Mindblindness: an essay on autism and theory of mind.* A Bradford Book, The MIT Press, 1995.
2. P. R. Butcher, A. F. Kalverboer, and R. H. Geuze. Infants' shifts of gaze from a central to a peripheral stimulus: A longitudinal study of development between 6 and 26 weeks. *Infant Behavior and Development*, 23:3–21, 2000.
3. G. E. Butterworth and N. Jarrett. What minds have in common in space: Spatial mechanisms serving joint visual attention in infancy. *British Journal of Developmental Psychology*, 9:55–72, 1991.
4. G. O. Deák, R. Flom, and A. D. Pick. Perceptual and motivational factors affecting joint visual attention in 12- and 18-month-olds. *Developmental Psychology*, 36:511–523, 2000.
5. J. L. Elman, E. A. Bates, M. H. Johnson, A. Karmiloff-Smith, D. Parisi, and K. Plunkett. *Rethinking Innateness.* A Bradford Book, The MIT Press, 1996.
6. I. Fasel, G. O. Deák, J. Triesch, and J. Movellan. Combining embodied models and empirical research for understanding the development of shared attention. In A. Jacobs, editor, *Intl. Conf. on Development and Learning*, pages 21–27. IEEE Computer Society, 2002.
7. J.M. Findlay and R. Walker. A model of saccade generation based on parallel processing and competitive inhibition. *Behavioral and Brain Sciences*, 22:661–674, 1999.
8. M. F. Land, N. Mennie, and J. Rusted. Eye movements and the roles of vision in activities of daily living. *Perception*, 28:1311–1328, 1999.
9. S. Leekam, S. Baron-Cohen, D. Perret, M. Milders, and S. Brown. Eye-direction detection: A dissociation between geometric and joint attention skills in autism. *British Journal of Developmental Psychology*, 15:77–95, 1997.
10. Chris Moore. Theories of mind in infancy. *British Journal of Developmental Psychology*, 14:19–40, 1996.
11. Brian Scassellati. Imitation and mechanisms of joint attention. In C. Nehaniv, editor, *Computation for Metaphors, Analogy, and Agents*, volume 1562 of *Lecture Notes in Artificial Intelligence.* Springer-Verlag, 1999.
12. W. Schultz, P. Dayan, and P. R. Montague. A neural substrate of prediction and reward. *Science*, 275:1593–1599, 1997.
13. R. S. Sutton and A. G. Barto. *Reinforcement Learning: an introduction.* A Bradford Book, The MIT Press, 1998.
14. E. Thelen and L. B. Smith. *A dynamics systems approach to the development of cognition and action.* A Bradford Book, The MIT Press, 1994.

CONNECTIONIST MODELS OF OVER-SPECIFIC LEARNING IN AUTISM

JOSEPH P. LEVY
School of Psychology and Therapeutic Studies,
Roehampton University of Surrey,
Whitelands College, West Hill, London SW15 3SN UK
E-mail: j.levy@roehampton.ac.uk

Autism is a complex neurodevelopmental condition causing difficulties in social interaction, communication, behavioural flexibility and superior performance in certain other abilities. Connectionist models are potentially useful tools to explore the differences in the developmental trajectories of normal and autistic children and adults. Several models are reviewed and a suggestion for a synthesis is suggested within a biologically constrained connectionist framework.

1. Introduction

Autism is a complex and pervasive condition affecting child development. It is primarily diagnosed by observing the presence of abnormal behaviour in three categories – the so-called "triad": social interaction, communication and behavioural flexibility. There is a wide degree of variation in the condition and its characteristics change as children develop into autistic adults. Autistic disorders are now often described as a "spectrum" from severely disabled individuals to those who can live independent lives but retain many autistic characteristics. It has been estimated that as many as 0.9% of children can be placed on the autistic spectrum.

The diagnostic spectrum of problems include those with social interaction such as the apparent inability to appreciate another person's point of view, problems with communication which usually lead to deficits or at least delays in language development and with behavioural flexibility which often leads to repetitive and stereotypical behaviours and a dislike of change. The understanding of autism is seen as a profound goal for psychology and cognitive science since it appears to be a condition that can tell us about the human ability to think about our own and other people's thoughts and be self-conscious. An up-to-date and accessible review of the current literature can be found in [1].

As well as the high-level behavioural patterns found in the triad, autism is always accompanied by lower-level characteristics, many of which allow autistic people to show better than average performance in isolated domains. These behaviours can usually be described in terms of a concentration on detail and individual features rather than wholes or gestalts. This bias toward local detail at the expense of the high level story may lead to enhanced performance in verbatim recall rather than the recall of gist, advantages in visuo-spatial tasks such as block design or the embedded figures test, a failure to be taken in by visual illusions [2], an ability to name isolated musical tones akin to perfect pitch [3] and even the development of savant skills in up to 10% of autistic people in areas such as drawing, music and calendar calculation.

There are no theories that successfully explain the whole range of behavioural characteristics in autism. However, I believe that the ability of connectionist models to learn and instantiate putative biological causes for developmental disorders make them a vital tool in the theoretical armoury for understanding autism. This paper takes a *neuroconstructivist* [4] view of development in accepting that any theory of development or developmental deficit has to be able to explain the dynamics of the trajectory through which the child's cognitive system moves.

2. Theoretical perspectives

There are many different theoretical accounts of autism (see [1]) but few make an attempt to explain all of its characteristics, instead usually concentrating on one or more of the diagnostic triad or the accompanying bias towards detailed processing. One class of theories makes the claim that autism amounts to a weakness in acquiring a "theory of mind" – the implicit knowledge that allows an individual to "mentalise" or represent the mental state of another person and so understand that person's perspective. Another position is that autism is caused by weakness in attentional or executive abilities causing a tendency to have rigidly fixed concentration on restricted areas of interest. Baron-Cohen [5] has recently claimed that there are two kinds of cognitive styles – a "male brain" type, which is driven to understand rule-based systems ("systemising") and a "female brain" type, which has a bias towards empathising. Most people have a balanced mixture of the two styles but it is claimed that autism is an example of an extreme male brain style.

The theory that appears to have inspired most computational modelling of autism is that of *weak central coherence.* Here, explanation is focused on those abilities outside the diagnostic triad where there appears to be a tendency to process and remember detail rather than gist and where there is a bias against integrating contextual information. Happé has suggested that central coherence is a cognitive style and so varies across the whole population [6].

3. Connectionist models

Connectionist models allow the exploration of ideas about the differences between learning in normal children and those with autism. Differences in starting conditions or constraints in the way that weights change can be used to observe departures from normal developmental trajectories. Until recently, most such models concentrated on general learning mechanisms that can be related to notions of weak central coherence. A notable exception is the paper in this volume by Triesch and Carlson [7] that discusses a model of the development of gaze control.

3.1. *Cohen (1994)*

Cohen [8] demonstrated how a simple network trained using the back-propagation algorithm could be manipulated in a way that mimics some of the general characteristics of autism. Cohen suggests that autistic children may have either too few neurons or too many in different parts of the brain. He suggests too many neurons in regions like the amygdala and hippocampus and too few in areas such as the cerebellum.

In his model, he either uses too few or too many hidden units while the network learns to discriminate between two groups of patterns. Too few hidden units produced a network that didn't discriminate the two groups in the test set well and so also performed poorly in discriminating previously unseen data in a test set. This corresponds to generally poor discrimination and performance. Too many hidden units produced excellent performance on the training set but, due to overfitting to accidental aspects of those specific items, generalised poorly on the test set. This corresponds to a pattern of performance that Cohen likened to the preference to specific detail that some autistic children show (e.g., in drawing, visuo-spatial skills, block design, precise echoing of intonation patterns, selective attention) and poor generalisation.

Cohen suggested that for higher performing children with too many connections, problems would not become apparent until overtraining was starting to occur and perhaps this explains why autism may only become apparent around 18-30 months. He suggests that early and intensive behavioural therapy may help by giving enough general experience to reinforce general patterns and prune out interfering overly specific ones.

3.2. *Gustafsson (1996)*

Another kind of neural network method for classifying patterns into groups is the self-organising feature map [9]. It has been hypothesised that this kind of mechanism is used in the development of topographical cortical maps. The method is unsupervised and can group together items classified with respect to many dimensions by projecting them onto a lower-dimensional, typically 2-D map.

Gustafsson [10] proposes that overly specific (narrow) cortical feature maps are what cause the memory-based and higher cognitive characteristics of autism. He suggests that good perceptual discrimination and a focus on local features are caused by narrow maps and that this "raw data" memory property could be a bottom-up explanation for the diagnostic triad of autistic behaviours – consequences of a general failure to extract salience by use of context.

The narrow selectivity is claimed to be due to "excessive lateral feedback inhibitory synaptic strengths". The feature maps may just be very poor and not achieve any kind of grouping or classification or they might be too narrow and so classify two similar items into different classes. These impairments then have knock-on effects for further cognitive integration as cognition continues to develop beyond primary feature maps.

Gustafsson suggests that remediation would depend on early diagnosis and concentration on developing the lowest order cortical maps to give a firmer foundation for the processing hierarchy above.

In work that can be related to Gustafsson's, Oliver, Johnson, Karmiloff-Smith and Pennington [11] describe a theoretical framework for studying developmental disorders, stressing how small initial differences in the parameters of a developing self-organising neural network can cause differences in outcome and how different starting points can lead to similar outcomes (relevant to the seemingly varied causes of autism). Simulations add support to the neuroconstructivist framework which claims that developmental disorders are best understood by considering how brain systems *develop* and that comparisons of developmental disorders to disorders caused by injuries to the ("end state") developed brain can be misleading.

They use a cortical matrix model with claims of biological plausibility. The networks are more biologically detailed than Gustaffson's model. The simulations concentrate on demonstrating how differences in developmental trajectories can be caused by altering different initial parameters. Mareschal &Thomas [12] suggest that the model could be a plausible implementation of Gustafsson's ideas of restrictive cortical maps being the root cause of autism. They contrast Gustafsson's lack of simulations but specific aim of explaining autism with Oliver et al.'s detailed but very general simulations. Both models are criticised for a failure to address high level behaviours.

Recently, this work on Kohonen networks has been extended to model the interaction between attention and unsupervised learning [13].

3.3. *McClelland (2000)*

McClelland [14] suggests that "hyperspecificity" as revealed by literal-mindedness, a lack of sensitivity to meaningfulness and good rote memory might be a possible representational cause for some autistic behaviour. He suggests that this may be caused by the excessive use of conjunctive coding in the representation of concepts in the autistic brain.

Referring to work that he and colleagues have done on memory and the interaction between cortex and hippocampus, McClelland describes the tension between generalisation and interference in the storage of memories. Overlap between patterns allows a network to generalise well but can cause interference.

One technique that reduces overlap is the use of conjunctive encoding, making use of specific combinations of features to store a memory. This can work well but causes a lack of generalisation, e.g., if a network remembers "red apple" and "green apple" as specific conjunctions of these features, it will later be unable to make use the generalisation that both objects were apples.

McClelland's suggestion is that the parts of autistic brains devoted to semantics and conceptual representations rely on conjunctive encoding too much and are thus unable to take advantage of generalisations but are very good at learning specific associations. The idea is related to weak central coherence and might be the consequence of mechanisms such as those suggested by Cohen and Gustafsson.

4. A synthesis

4.1. *LEABRA*

LEABRA (Local, Error-driven and Associative, Biologically Realistic Algorithm) [15] is a biologically constrained framework for the computational modelling of cognition. It allows fully recurrent activation flow with a k-winners-take-all competition between units within a layer. Learning is achieved by a combination of a Hebbian learning rule and a recurrent error-driven algorithm related to contrastive Hebbian learning or the deterministic Boltzmann machine. For full details of LEABRA see [16]. O'Reilly and Munakata have used the framework to model several different cognitive domains. The richness of the biologically constrained processing and learning primitives make the framework potentially attractive for cognitive modelling.

4.2. *Using LEABRA to model over-specific learning in autism*

As we have seen above, previous work on the computational modelling of autism has concentrated on representational or learning mechanisms that may account for over-specific learning in autism. I suggest that LEABRA can provide a synthesis of the three models reviewed above and that this may act as a springboard to more general models of specific phenomena within autism. LEABRA could be argued to be a connectionist cognitive architecture and as such allow different theorists to share a common and biologically constrained, if elaborate, framework so that different features and assumptions can be easily compared.

Since LEABRA encompasses both unsupervised and supervised forms of learning it can emulate Cohen's model of over-specificity due to over-training and Gustafsson's idea that excessive lateral inhibition can cause a self-organising learning algorithm to capture over-restrictive categories. However, I suggest that a better way to synthesise these ideas is to focus on McClelland's suggestion that excessive conjunctive feature representations are the end result of over-specific learning. There are several parameters within LEABRA, which could cause such an end result in the way that network dynamics, Hebbian learning and contrastive Hebbian learning interact.

LEABRA allows the balance between unsupervised and supervised influences on weight change to be varied. In simple learning tasks, a purely error-driven learning process can achieve good results but usually better and faster learning and better generalization are achieved by using a small proportion (1% or less) of unsupervised Hebbian learning on the weights. Changes in learning parameters that lead to differing amounts of conjunctive hidden unit representations and hence differing degrees of over-specific learning can be illustrated using an example from Chapter 4 of reference [16]: a three layer network identifies which vertical and horizontal lines are present on a 5x5 input retina using one output unit per possible line on a 10 unit output layer (5 possible horizontal lines and 5 possible vertical lines). A hidden layer contains 25 units.

Pure error-driven contrastive-Hebbian learning produces very distributed (and thus in a sense highly conjunctive) representations in a hidden layer. This can be seen by looking at Figure 1 where (following the exercise in O'Reilly & Munakata, 2000), the projection of weights from the input retina to each hidden unit is depicted. The greyscale rendering of this colour diagram doesn't show the different activations very distinctly but it is clear that many input units contribute to the activation of each hidden unit. Figure 2 shows what happens when a small proportion of Hebbian influence produces hidden unit representations that are more feature-like and far less conjunctive. If competition between units in a hidden layer is relaxed, more can become active during mixed supervised and unsupervised learning and the hidden units have an increased tendency to act as detectors for conjunctions of input information rather than single features and this over-specificity decreases the degree of generalisation to novel inputs after learning. This is illustrated in Figure 3 where the k parameter in the kWTA inhibitory function in LEABRA is increased from 6 to 12 during combined error-driven and Hebbian learning. Other ways of influencing the degree of distribution of activation over hidden units that results after learning are possible within LEABRA including making the weight gain parameter less sharp. The results of decreasing this parameter from 6 to 3 on the learned hidden unit representations can be seen in Figure 4.

The different ways of achieving over-specific representation can be related to some of the different models in the literature and may well give clues about different biological underpinnings that may result in behaviour that is classed within the heterogeneous diagnosis of autism. However, they also demonstrate that there are several possible ways of producing over-specific learning in even a simple network and more is required for an explanation of autism. LEABRA provides an opportunity to model these different developmental trajectories and ground them within biological constraints.

A further advantage in using LEABRA is that it has already been used to model many cognitive domains and these models can stand as tests for the consequences of any hypothesized changes in learning parameters that arise from more general speculation on the causes of autistic behaviours. An example might be that a change in a parameter that is suggested to capture over-specific learning and poor generalization could be applied to existing models of the development of low-level vision to see whether the characteristics of autistic people's visual systems can be captured.

5. Ways forward

The brief survey above demonstrates that some very general aspects of autism can be captured by simple connectionist principles. However, the manipulations made such as increasing the number of hidden units or increasing lateral inhibition could be said to be "free gifts" [17] and so predictable from past work that they hardly need to be simulated. To go beyond this, connectionist models of autism need to make closer contact with the data and operationalise autistic behaviours in a more satisfying way.

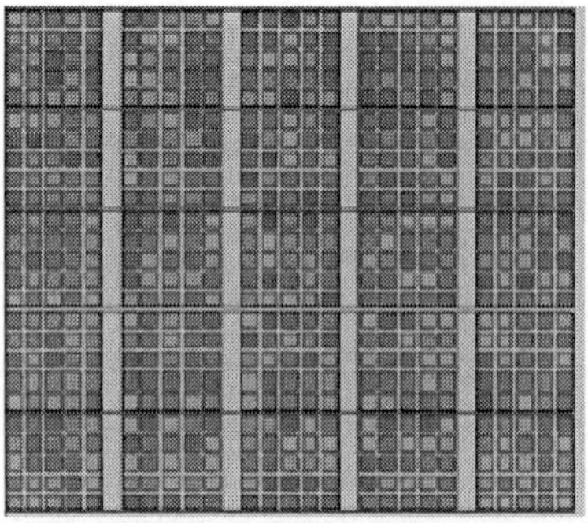

Figure 1. Projection of each input unit to each hidden unit after pure error-driven learning

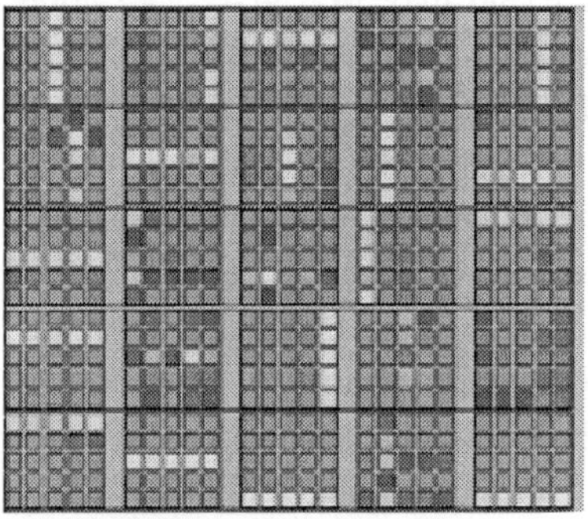

Figure 2. Hidden unit representation after combined error-driven and Hebbian learning.

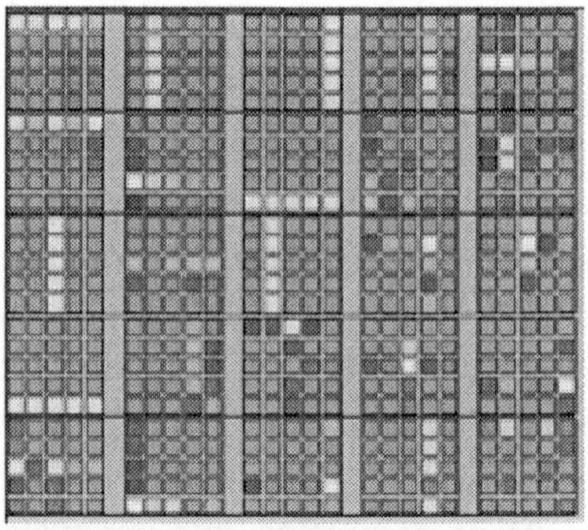

Figure 3. Hidden unit representation after decreasing hidden unit competition.

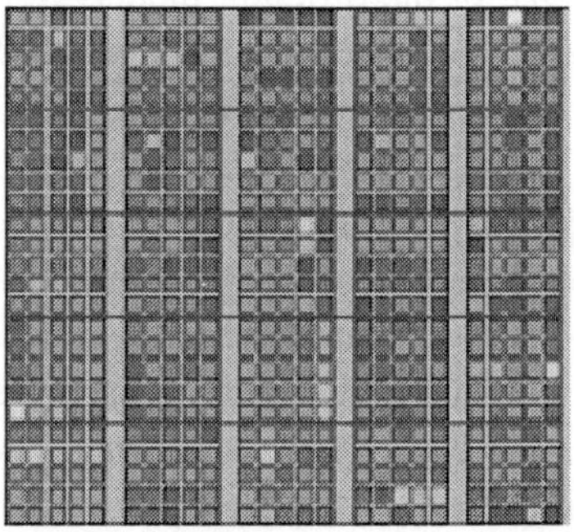

Figure 4. Hidden unit representations after decreasing weight gain parameter

I suggest that one way forward would be to examine parameters such as the ones described above in an established framework like LEABRA that can cause over-specific learning or change the balance between specificity and generalization and apply these models to several of the characteristics of autism. Weak central coherence based explanations of autism would be greatly strengthened if a computational model could be shown to capture data from experiments covering attentional and theory-of-mind tasks as well as tasks related to perceptual and linguistic coherence and the balance between local and global processing. Since the parameter changes are being made to a biologically constrained framework, they can be inspired by known facts of abnormalities in autistic brains. Equally, any successful computational manipulation can be checked against biological facts. Any successful models can be extended by applying them to existing models that have been implemented in LEABRA.

Acknowledgements

I would like to thank two anonymous reviewers for comments on a draft version of this chapter.

References

1. Frith, U. *Autism: Explaining the Enigma*, 2nd edn. Oxford: Blackwell. (2003)
2. Happe , F.G.E. Studying weak central coherence at low levels: Children with autism do not succumb to visual illusions: A research note. *Journal of Child Psychology and Psychiatry*, **37**, 873-877 (1996).
3. Heaton, P., Hermelin, B. and Pring, L. Autism and pitch processing: A precursor for savant musical ability? *Music Perception,* **15 (2)** 291-305 (1998).
4. Karmiloff-Smith, A. Development itself is the key to understanding developmental disorders.*Trends in Cogn. Sci.* **2**, 389-98 (1998).
5. Baron-Cohen, S. *The Essential Difference: Men, Women and the Extreme Male Brain*. London: Allen Lane. (2003)
6. Happe , F., Autism: Cognitive deficit or cognitive style? *Trends in Cogn. Sci.* **3(6)**, 216-222 (1999).
7. Triesch, J. and Carlson, E. *A Computational Model of the Emergence of Gaze Following*, chapter in this volume.
8. Cohen, I. L. An Artificial Neural network Analog of learning in Autism. *Biological Psychiatry*, **36(1)**, 5-20 (1994).
9. Kohonen, T., *Self-Organising Maps*, 3rd edn. Springer-Verlag, Berlin (2001).
10. Gustafsson, L. Inadequate Cortical feature maps: A Neural Circuit Theory of Autism. *Biological Psychiatry*, **42**, 1138-1147 (1997).
11. Oliver, A., Johnson, M. H., Karmiloff-Smith, A. & Pennington, B. Deviations in the emergence of representations: A neuroconstructivist framework for analyzing developmental disorders. *Developmental Science*, **3:1**, 1-40 (2000).
12. Mareschal, D. & Thomas, M. S. C. *Self-organization in normal and abnormal cognitive development*, in Kalveboer, A. F. & Gramsbergen, A. (eds) Brain and Behaviour in Human development: A source book. Kluwer. (2000).

13. Gustafsson, L. and Paplinski, A. P. *Preoccupation with a Restricted Pattern of Interest in Modelling Autistic Learning,* Paper presented at KES'2003 Seventh International Conference on Knowledge-Based Intelligent Information & Engineering Systems Oxford, UK. (2003)
14. McClelland, J. L. The basis of hyperspecificity in autism: a preliminary suggestion based on properties of neural nets. *Journal Of Autism And Developmental Disorders*, **30(5)**, 497-502 (2000).
15. O'Reilly, R. C. Six principles for biologically-based computational models of cortical cognition. *Trends Cogn. Sci.*, **2**, 455-462 (1988).
16. O'Reilly, R.C. & Munakata, Y. *Computational Explorations in Cognitive Neuroscience: Understanding the Mind by Simulating the Brain.* Cambridge, MA: MIT Press (2000).
17. Bullinaria, J.A. Free Gifts from Connectionist Modelling. In R. Baddeley, P. Hancock & P. Földiák (Eds), *Information Theory and the Brain,* 221-240. Cambridge: Cambridge University Press (2000).

Category Acquisition

SOLVING THE VISUAL EXPERTISE MYSTERY

C. A. JOYCE AND G. W. COTTRELL

University of California, San Diego
Computer Science and Engineering
9500 Gilman Dr.,
La Jolla, CA 92093-0114, USA
E-mail: {gary,cjoyce}@cs.ucsd.edu

Through brain imaging studies and studies of brain-lesioned patients with face or object recognition deficits, the fusiform face area (FFA) has been identified as a face-specific processing area. Recent work, however, illustrates that the FFA is also responsive to a wide variety of non-face objects if levels of discrimination and expertise are controlled. The mystery is why an expertise area, whose initial domain of expertise is presumably faces, would be recruited for these other domains. Here we show that features tuned for fine-level discrimination within one visually homogeneous class have high-variance responses across that class. This variability generalizes to other homogenous classes, providing a foothold for learning.

1. Introduction

There has been a great deal of progress in understanding how complex objects, in particular, human faces, are processed by the cortex. However, there is also controversy about the roles of various cortical areas, especially the Fusiform Face Area (FFA).[1,2,3] Is the FFA a "module," specific to the domain of faces, or is it instead specific to the process of fine level discrimination? Damage to the FFA leads to prosopagnosia[4] (the inability to recognize faces), but it is unclear how face-specific this processing deficit is. Further, some researchers have shown, using fMRI, that when the level of expertise is controlled, the FFA is activated in car, bird, and Greeble (a class of fictional objects, see Figure 1, right column) experts when they view their respective categories of expertise.[5,6,7] This suggests that the FFA is a *fine level discrimination area*. The issue we address in this paper is why an area that presumably starts life as a face processing area (this being the first domain of expertise) is *recruited* for these other types of stimuli?

In addressing this question, the definition of "expertise" is critical. We adopt Gauthier and Tarr's operational definition of the term: someone is an expert if they are as fast to identify members of a category as individuals (subordinate level) as they are to verify their category membership (basic level). For example, a bird expert would be as fast/accurate at identifying a picture of a bird as an "Indigo Bunting" as at identifying it as a "bird." When training a subject in a novel category, the convergence in reaction times in these two tasks is called the "entry level shift."

This study replicates and expands on previous work[8] in which we have shown that neurocomputational models trained to make fine level discriminations learn individuation of Greebles faster than models that have not been trained to become experts in any domain. This suggests that, if there is a *competition* between cortical areas to solve tasks, as has been suggested previously,[9,10] the FFA would be primed to win the competition for a novel expertise task. Here, we show why this happens.

2. Experimental Methods

To investigate this issue, neural networks were trained on Greeble identification following various pretraining regimens.

The stimulus set consisted of 300 64x64 8-bit grayscale images of human faces, books, cans, cups, and Greebles (60 images per class, 5 images of 12 individuals, see Figure 1). The five images of each indivudual within each category were created by randomly moving the item 1 pixel in the vertical/horizontal plane, and rotating up to +/-3 degrees in the image plane.

Images were preprocessed by applying Gabor wavelet filters as a simple model of complex cell responses in visual cortex, extracting the magnitudes (which makes them nonlinear and somewhat translation invariant), normalizing via z-scoring, and reducing dimensionality to 40 via principal component analysis (PCA)[11]. Greeble images were not used to generate the principal components in order to model subjects' lack of experience with this category.

A standard feed-forward neural network architecture (40 input units, 60 standard logistic-sigmoid hidden units, variable numbers of linear output units) was used. Networks were trained using a learning rate of 0.01 and momentum of .5.

During pretraining, *all* networks (basic and expert) learned to perform basic level categorization on all 4 non-Greeble categories. *Expert* networks were additionally taught to perform subordinate level categorization of one of the four categories. Basic level networks had 4 output nodes correspond-

Figure 1. Example stimuli

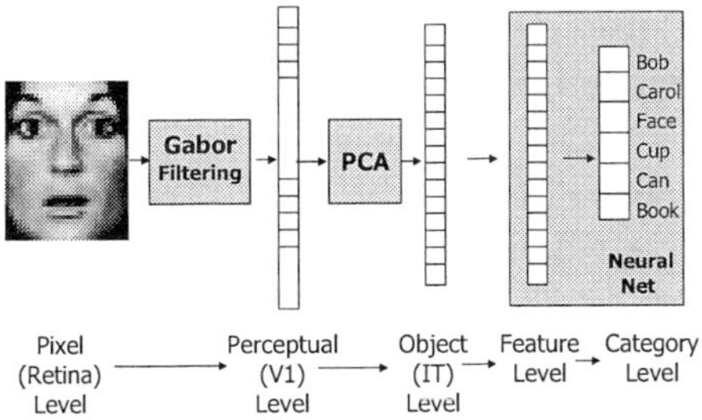

Figure 2. The expertise model. The feature level is where task specific features are developed and variance is measured in Figure 7.

ing to book, can, cup, and face. Expert networks had 14 outputs: 4 for the basic categories, and 1 for each of the 10 individuals (e.g. can1, can2, ... can10, for a can expert). In phase two, the pretrained networks learned subordinate level Greeble categorization along with their original task. Eleven output nodes were added: 1 for the basic level Greeble categorization, and 1 for each Greeble individual. The network then performed a 15-way (basic network) or 25-way (expert network) classification task. All networks were trained on 30 images (3 images of 10 individuals) per class during pretraining and 30 more images of Greebles in phase 2. Thus any differences in representation are due to the task, not experience with exemplars. To test for generalization, 29 images were used (one new image of each of the expert category individuals (10 + 10), plus 3 images of novel basic level exemplars per category).

Ten networks, each with different random initial weights, were trained on each of the 5 pretraining tasks (basic, or face/can/cup/book expert) for 5120 epochs. Image sets were randomized. Intermediate weights of each network were stored every $5 * 2^n$ epochs, for n=1:10. Phase 2 training was performed at each of these points ("copying" the network at that point) to

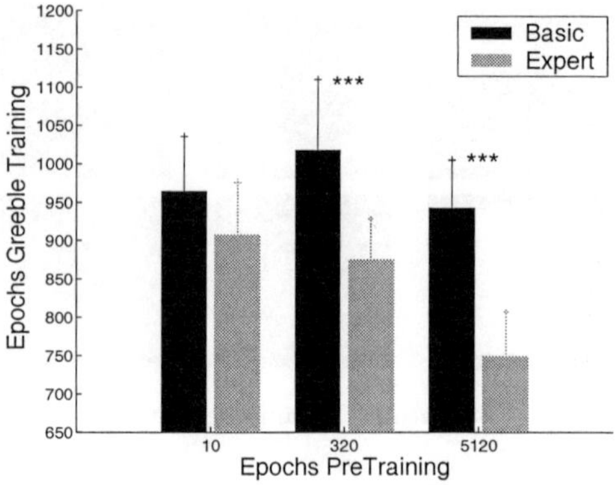

Figure 3. Number of epochs to learn the new task based on number of pretraining epochs. Error bars denote +/-1 standard deviation.

observe the time course of expertise effects. Training concluded when the RMSE of the Greebles fell below .05. Thus, there were a total of 550 phase 2 networks.

3. Results

All networks reached an RMSE of less than .0012 by the completion of 5120 pretraining epochs, with basic networks learning faster than expert networks.

Figure 3 shows the average number of epochs required for networks of each type to learn the subordinate Greeble task at three levels of pretraining epochs. The basic level networks took by far the longest to learn the Greeble task, obtaining no benefit from more pretraining cycles. All of the expert networks learned the Greeble task significantly faster if they were given more pretraining on their initial expert task, with faces benefitting the most from additional pretraining (data not shown).

3.1. *Entry Level Shift*

Training paradigms with human subjects use the reaction time entry level shift to determine a subject's expert status. Example data from a human Greeble expert is shown in Figure 4a. In networks, reaction time is modelled as the amount of uncertainty in the output of the network. This uncertainty is measured by taking 1 minus the logistic of the output activation on the node corresponding to the correct category or individual classification for each output pattern. Figure 4b shows the Greeble entry level shift for a network pretrained as a book expert. Note that response time to subordinate level classification of books is as fast as basic level classification prior to Greeble training.

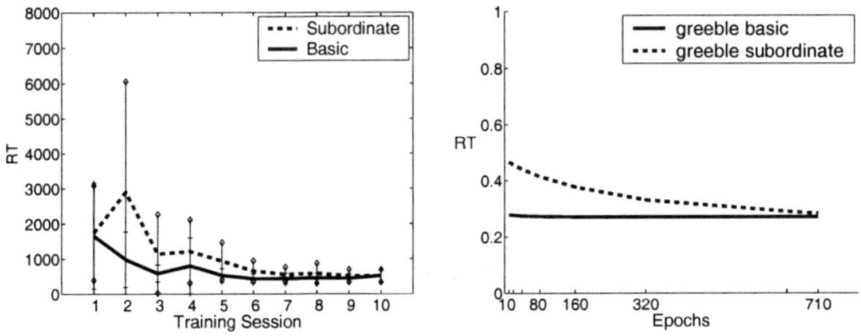

(a) (b)

Figure 4. Entry level shift for the Greeble task. (a) Human data from one of our experts. (b) Network data.

3.2. *Network Plasticity*

In previous work[8], we hypothesized that the hidden units in the expert networks would tend to stay in the linear range, in order to better perform the fine level discrimination task. We suggested that this would lead to faster learning of the new task, since the higher slope of the hidden units would result in faster weight changes. The slope of the hidden units has been called a measure of plasticity in previous work.[12]

Plasticity to a stimulus category is measured as the average value of the slope of the activation function (here the logistic sigmoid) across all hidden units for all input patterns from that category. Unexpectedly, results indicated that lower plasticity networks learned the new task faster. Figure 5 shows the plasticity of the pretrained networks in response to the stimuli used during pretraining (left), and to the new set of untrained Greeble patterns (right). For all patterns (pretrained and novel), non-expert networks retained their plasticity better across pretraining epochs than experts. Furthermore, plasticity to the (untrained) Greebles *decreased* over training on the expert task. This paradox may be resolved in part if the plasticity measure is viewed as a measure of *mismatch* between the stimuli and the weight vectors – the closer the weight vectors line up with the stimuli (either in

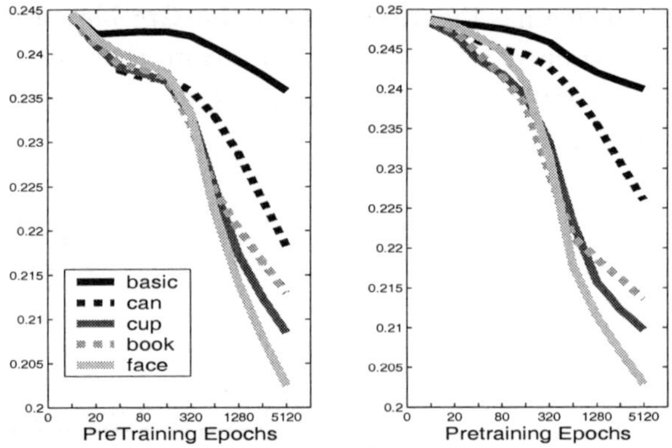

Figure 5. Average plasticity of the hidden units over training to learned categories (left), and novel Greebles (right).

the same or opposite direction), the more the hidden units will be activated or inactivated. Thus, here the weight vectors are simply becoming more aligned with the stimuli, and, perhaps surprisingly, also more aligned with the Greeble images. This is not the whole story, however, as we will see in the next section.

3.3. *Hidden Unit Activation*

Since expert network representations become less plastic with training, how does the network actually discriminate one individual from another within and across categories?

The activation of the hidden units in response to each category of stimulus provides some explanation. Figure 6 shows the activation levels of 3 representative hidden units from a basic level (a,b) and a face expert (c,d) network in response to individual training patterns both prior to (column 1) and after (column 2) Greeble training.

Prior to Greeble training (column 1), the hidden units in subordinate level networks (Figure 6c) show more variability of response across input patterns than do basic level networks (Figure 6a). After Greeble training, both basic and expert level networks show more variability in hidden unit activation across input patterns (Figure 6b,d).

These results suggest that correct discrimination requires a representation that is distributed across multiple hidden units which modulate in different ways in response to different input patterns from the same class.

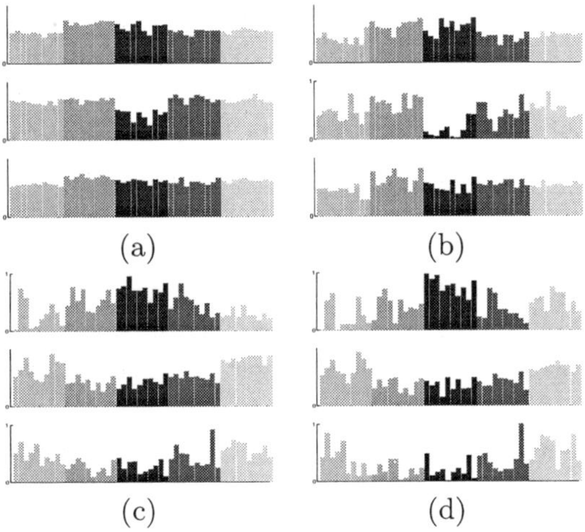

Figure 6. Single unit recordings from networks for face, book, can, cup, and greeble patterns, respectively. a) basic network, pre-Greeble training; b) basic network, post-Greeble training; c) face expert, pre-Greeble training; d) face expert, post-Greeble training.

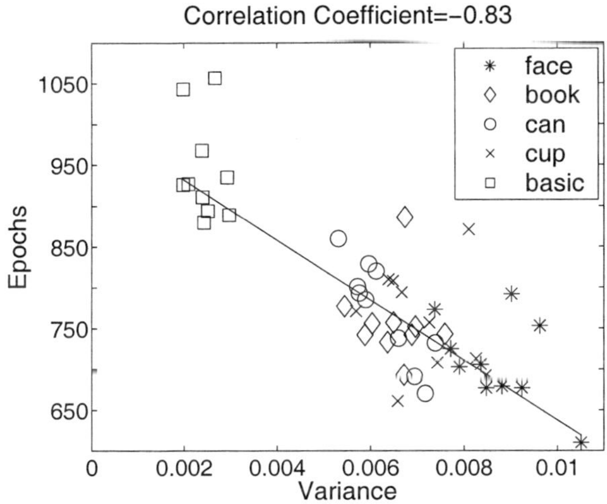

Figure 7. A regression of Greeble pre-training variance versus training time.

3.4. *Relationship of Variability to Learning*

There appears to be a provocative relationship between learning and hidden unit variability: networks that have learned a subordinate level task and exhibit hidden unit variability, also learn a secondary subordinate level task faster than basic level networks which exhibit little hidden unit variability. This suggests two things: 1) variability should increase with experience, particularly when making a subordinate level discrimination, and 2) the amount of variance a network exhibits in response to a category prior to training on that category should be predictive of the speed with which that category is learned. The first hypothesis is addressed by examining how variability changed over the course of pretraining: 1) variability increases for all categories in all networks as the number of training epochs increases; 2) increases in variability are much larger for expert networks than basic networks, and are largest for the category being learned at the subordinate level; 3) expert networks show more variability to all categories than basic networks, even to categories being learned at the basic level; 4) even variability to Greebles, which the network has never been trained on in any manner, increases with pretraining epochs, although not as much as the categories being trained (at both subordinate and basic levels). These results support the conclusion that pretraining causes networks, particularly those making a subordinate level discrimination, to learn features which generalize well to new categories.

Figure 7 illustrates the second hypothesis: that amount of variability to Greebles, prior to training on them (x-axis), should be predictive of how fast the network can learn the Greeble task (y-axis). There is a strong negative linear correlation between these two variables for expert networks such that those exhibiting the lowest variance also take the longest to learn the Greeble task ($r^2 = -.53, p < .001$). For basic networks, there is no significant correlation between variance and learning time ($r^2 = -.21, p = .557$). Those expert networks exhibiting the highest variance and lowest Greeble learning time, are the networks that initially learned faces, the task that was the most difficult to learn in pretraining. This suggests a relationship between the difficulty of the pretraining task, and the ease with which subsequent subordinate discriminations can be learned.

4. Conclusions

The results of these simulations are indicative of a system in which expertise results from the flexible use of fine-tuned feature representations. Further, the types of features learned through subordinate level discrimination of visually different categories seem to generalize well to new categories. Finally, learning difficult perceptual discriminations enables faster learning of new discriminations. These results suggest that the FFA fine–tunes its sensitivity to small differences in homogeneous stimuli when given a novel, fine–level discrimination task.

It might be considered counter-intuitive that an expert network with low plasticity at the hidden layer should yield more variable responses across hidden units. The measures themselves explain how this can occur. Maximum plasticity occurs when there is a large mismatch between inputs and weights (i.e., they are orthogonal). As the network becomes more expert, the inputs and weights become more similar/matched (i.e., it loses plasticity). Basically, the weights become more tuned to the specific input vectors and, for expert networks, more responsive to the small differences between them. Thus, the resulting hidden unit activations become more variable because they correspond more closely to the fine-level differences between the input patterns (for the expert networks).

A critical question, then, is what exactly are the features the FFA uses? More research is required to address this question, but clearly these features must be broad enough to encompass vastly visually different stimuli. In further work we will investigate the possibility that these features result from combinations of lower level visual sensitivities of the cells that feed into FFA - for example, cells which are sensitive to low spatial frequencies. Thus the features coded in this area could be reflections of early, lower-level visual processing biases.

Acknowledgments

This work was supported by the McDonnell Foundation (Perceptual Expertise Network, 15573-S6) and NIMH MH57075 grant to GWC.

References

1. N. Kanwisher, J. McDermott, and M. M. Chun. The fusiform face area: A module in human extrastriate cortex specialized for face perception. *J Neurosci*, 17:4302–4311, 1997.
2. N. Kanwisher. Domain specificity in face perception. *Nat Neurosci*, 3 (8):759–763, August 2000.
3. M. J. Tarr and I. Gauthier. Ffa: A flexible fusiform area for subordinate-level visual processing automatized by expertise. *Nat Neurosci*, 3(8):764–769, August 2000.
4. E. De Renzi, D. Perani, G.A. Carlesimo, M.C. Silveri, and F. Fazio. Prosopagnosia can be associated with damage confined to the right

hemisphere — An MRI and PET study and a review of the literature. *Psychologia*, 32(8):893–902, 1994.
5. I. Gauthier and M.J. Tarr. Becoming a "greeble" expert: Exploring mechanisms for face recognition. *Vision Res*, 37(12):1673–1682, 1997.
6. I. Gauthier, M. J. Tarr, A. W. Anderson, P. Skudlarski, and J. C. Gore. Activation of the middle fusiform "face area" increases with expertise in recognizing novel objects. *Nat Neurosci*, 2(6):568–573, June 1999.
7. I. Gauthier, P. Skudlarski, J. C. Gore, and A. W. Anderson. Expertise for cars and birds recruits brain areas involved in face recognition. *Nat Neurosci*, 3(2):191–197, 2000.
8. Maki Sugimoto and Garrison W. Cottrell. Visual expertise is a general skill. In *Proceedings of the 23rd Annual Conference of the Cognitive Science Society*, Mahwah, New Jersey, 2001. Lawrence Erlbaum Associates.
9. G.M. Edelman. *Neural Darwinism: The theory of neuronal group selection.* Basic Books, Inc., New York, NY, 1987.
10. Robert A. Jacobs, Michael I. Jordan, Steven J. Nowlan, and Geoffrey E. Hinton. Adaptive mixtures of local experts. *Neural Comput*, 3:79–87, 1991.
11. M. N. Dailey and G. W. Cottrell. Organization of face and object recognition in modular neural network models. *Neural Netw*, 12(7–8): 1053–1074, 1999.
12. A.W. Ellis and M.A. Lambon Ralph. Age of acquisition effects in adult lexical processing reflect loss of plasticity in maturing systems: Insights from connectionist networks. *J Exp Psychol Learn Mem Cogn*, 26(5): 1103–1123, 2000.

EMPIRICAL EVIDENCE AND THEORETICAL ANALYSIS OF FEATURE CREATION DURING CATEGORY ACQUISITION

M. FINK AND G. BEN-SHAKHAR

Interdisciplinary Center for Neural Computation and Psychology Department,
The Hebrew University of Jerusalem,
Mt. Scopus, Jerusalem 91905, Israel
E-mail: fink@huji.ac.il

D. HORN

School of Physics and Astronomy, Tel Aviv University
Ramat Aviv, Tel Aviv 69978, Israel
E-mail: horn@post.tau.ac.il

This study is aimed at detecting factors influencing perceptual feature creation. By teaching several new perceptual categories, we demonstrate the emergence of new internal representations. We focus on contrasting the role of two basic factors that govern feature creation: the informative value and the degree of parsimony of the feature set. Two methods of exploring the structure of internal features are developed using an artificial neural network. These methods were empirically implemented in two experiments, both demonstrating a preference for parsimonious internal representations, even at the expense of the informative value of the feature. Our results suggest that feature parsimony is maintained not only to optimize the resource management of the perceptual system but also to aid future category learning.

1. Introduction

In our daily lives we recognize elaborate perceptual categories with remarkable speed and accuracy [1], often learning to detect new categories after single exposures to exemplars [2]. In order to explain these sophisticated capabilities it has been proposed that classical feature based theories [3] should be extended by emphasizing the perceptual system's capacity to create new complex features [4]. Previous experiments have shown that features are created both by extraction of input statistics in unsupervised settings [5], and by providing feedback during category learning [6]. However, the specific factors underlying feature creation during new category acquisition have not been elaborated, and are therefore the focus of our research. One method that has been suggested for feature-selection emphasizes maximization of the information supplied by the features on the required categories [7]. We hypothesize that feature information value is not the only criterion for feature selection, and highlight the importance of feature parsimony, as a second dominant factor in this process. We regard parsimony as a minimization of input elements required for feature activation.

2. A Model Problem for Contrasting Information and Parsimony

We start by defining a model problem that will enable us to contrast the roles of information and parsimony as factors in feature creation. Our model problem consists of eight dimensional input elements ($p_{i,i=1,..,8}$). Each binary input element could be in an *on* or *off* state (p_i={1,-1})[a]. The target output includes four binary actions ($t_{i,i=1,..,4}$). The system is required to learn four categories ($C_{i,i=1,..,4}$). These four categories are defined as mappings from the input set {**p**} to the target-actions {**t**}. Each category is defined by four specific input elements in an *on* position associated with an activation of a single target element (see table 1 and [8] for input description).

Table 1: A description of the four categories to be learned (for each category 1 and -1 indicate which elements must be in an *on* or *off* state, respectively and * denotes the category's indifference to a certain element).

	Input elements									Target elements			
	1	2	3	4	5	6	7	8		1	2	3	4
Category 1	1	1	1	1	*	*	*	*	→	1	*	*	*
Category 2	*	*	1	1	1	1	*	*	→	*	1	*	*
Category 3	*	*	*	*	1	1	1	1	→	*	*	1	*
Category 4	1	1	*	*	*	*	1	1	→	*	*	*	1
Otherwise	*	*	*	*	*	*	*	*	→	-1	-1	-1	-1

2.1. *Feature Structure of the Model Problem*

We limit our analysis to two possible solutions for the model problem, one based on quadruple detectors and one on pair-features. Quadruple detector features emerge directly from the category structure described in table 1 (each category requires four specific input elements to be in an on state while being indifferent to the remaining four input elements). However, a second possible feature set might be used to solve the model problem. As shown in table 1, categories 1 and 2 share two common required input elements (p3 & p4). In fact, each category shares two pairs of required input elements with two other categories. Thus, we can redefine the input of each category as a conjunction of two pairs (Figure 1).

We will now try to justify our decision to limit the analysis to two possible feature sets, while relating each of these options to a different dominant factor in

---

[a] We select a [1,-1] activation semantics to indicate the presence and absence of a certain feature. Thus, 0 represents a natural selection of a neutral state. The reduced intensity of "imagery" activation will later be represented by the value ½.

the feature creation process. The advantage of quadruple detectors over all other feature sets emerges from analyzing the mutual information the various features share with the required categories. Mutual information measures the reduction in category uncertainty (entropy) when certain feature detectors are available. An examination of all 256 features that require certain inputs to be active $\{1,*\}^8$, reveals that under the exemplar distribution, quadruple detectors provide the maximum information for category detection. The maximum information hypothesis [7] states that features should be selected by their information value. Therefore, implementing this hypothesis in our model problem will lead to a selection of the quadruple detector set. In contrast a (slightly less informative) pair feature representation (**h**) reduces the feature set's complexity by only requiring conjunctions of the second degree. We therefore regard the pair-features as a manifestation of a preference for parsimonious representations.

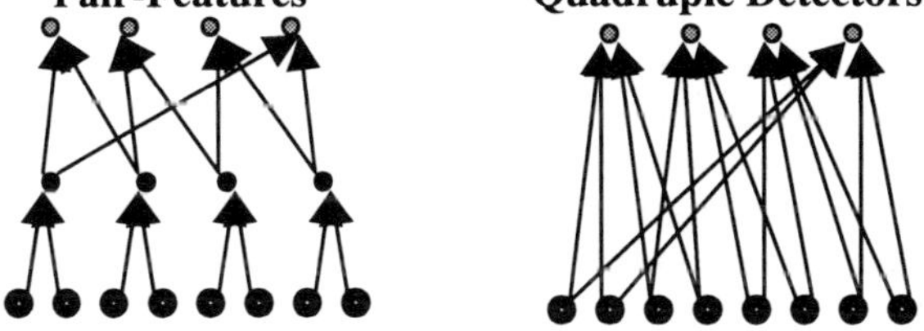

pair-feature:	$\mathbf{h}_1$	$\mathbf{h}_2$	$\mathbf{h}_3$	$\mathbf{h}_4$
inputs in an *on* position	1 & 2	3 & 4	5 & 6	7 & 8

Figure 1: A description of the quadruple detectors and the pair-features solutions

It is important to notice that the model problem was designed to demonstrate emergence of hierarchical structure in the simplest possible framework (two existing features combine to create a new complex feature). In this sense we view the model problem as a step in an induction process. In this induction process the input of the model problem should be regarded as an output of previously trained feature detectors and in a similar fashion the output units should be regarded as candidate features for more complex future categories[b].

---

[b] Since the model problem is viewed as a feature creation induction step, the same activation range [-1,1] was selected for input units, internal features and output units.

2.2. *Empirical Implementation of the Model Problem*

In order to test the emerging representation we implemented the abstract model problem in a concrete experimental setting. The eight-dimensional binary inputs were implemented by pictures composed of eight colored cubes ($p_{i,i=1,..,8}$). For each cube one color was selected to function as the *on* state and another color as the *off* state ($p_i=\{1,-1\}$). Thus a total of 16 colors were used (Figure 2).

i	*on*	*off*	$p_i[1]$	$p_i[2]$
1	black	pink	-1	-1
2	blue	peach	1	-1
3	green	light blue	1	1
4	yellow	gray	1	1
5	orange	beige	-1	-1
6	white	brown	1	-1
7	red	cyan	1	1
8	purple	light green	1	1

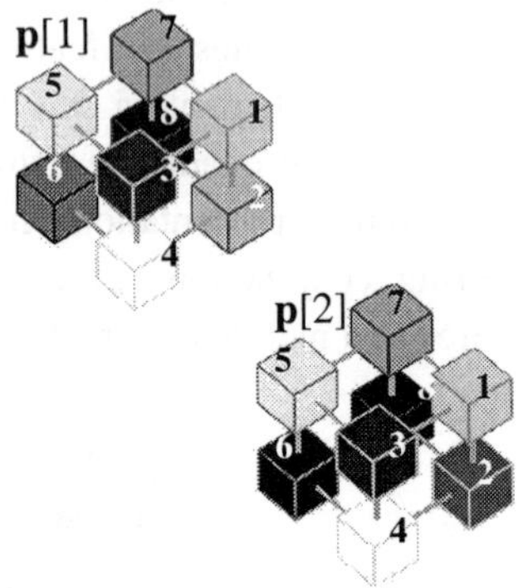

Figure 2: An example of two stimuli **p**[1] & **p**[2] composed of eight binary color-cubes each appearing in two randomly allocated distinct colors (presented in shades of gray). For simplicity the following figures will be presented in this view although participants viewed the pictures rotating so that all cubes had an equally salient visual positioning.

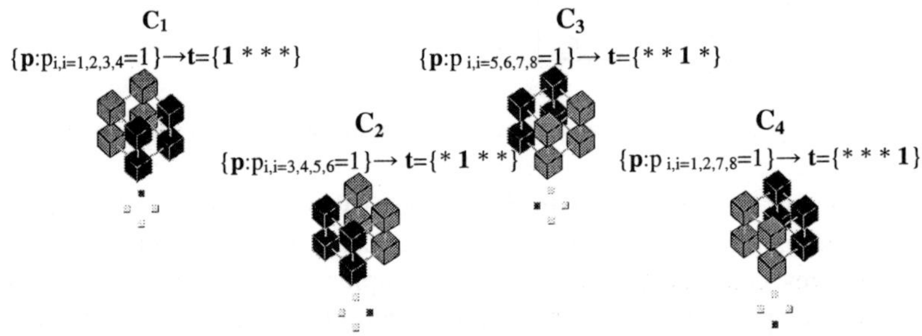

Figure 3: The four categories and target buttons learned by participants (e.g. $\mathbf{p}:p_{i=1,..,4}=1$ indicates that category 1 includes images in **p** where input elements 1 through 4 are in an *on* state). In this figure and in the following ones black indicates cubes required to be in the *on* state ($p_i=1$) for the category to be present in the picture. Gray indicates cubes that were irrelevant to the category and could be either in an *on* or *off* state ($p_i=\{1,-1\}$).

Beneath the pictures, an array of four target-buttons was presented ($t_{i,i=1,..,4}$). Four categories ($C_{i,i=1,..,4}$) were defined as a mapping from the picture set {**p**} to the target buttons {**t**}. Each category was based on four neighboring, cubes in an *on* position. Exemplars of each category were generated by using color combinations of the remaining four non-relevant cubes (Figure 3).

The intersections of the categories' relevant cubes defined the set of *pair-features* (see Figure 4). Each of these four *pair-features* is congruent with the requirements of two categories. It was previously hypothesized that an internal representation of these *pair-features* might evolve due to their parsimony and efficient representation of the target categories.

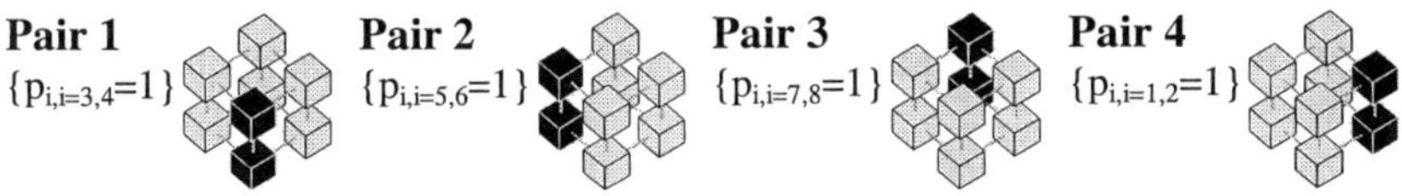

Figure 4: the *Pair-features* (black) formed by the intersections of the four categories.

The initial learning session was composed of four stages. At each stage, participants learned one additional category by trial-and-error. In every trial one picture was displayed and participants were required to press an appropriate target button. If a wrong button was pressed an error tone was activated. This procedure continued until a criterion of 100 consecutive successes was met, indicating that the participants had learned to associate the new category pictures with the designated target-button. Next, we tested whether the hypothesized parsimonious pair-features have emerged.

2.3. *A Neural Network Realization*

In order to examine whether the quadruple detectors or the pair-features have been created, methods for exploring the hidden structure of a perceptual system should be designed. A simple Multi Layer Perceptron Neural Network is used to demonstrate the proposed methods. This network contains:

- Eight input elements ($p_{i,i=1,..,8}$)
- Four hidden layer neurons ($h_{i,i=1,..,4}$)
- Four output neurons ($a_{i,i=1,..,4}$) each representing a possible action t_i

Although our model problem was defined using binary input and output elements, the simulation neurons used continuous [1 1] sigmoid transfer functions. The neurons of each layer were fully connected to the previous layer's elements by assigning a real number to represent the synaptic efficacy between the pre-synaptic and post-synaptic elements. The training set followed the distribution of eight-dimensional binary-inputs {**p**} paired with the appropriate four-dimensional binary output vectors {**t**}. Back propagation training was performed following [9]. A detailed description of the training process is available at [10]; nevertheless our focus is not on using the neural network as a model for the training process but rather as a tool for developing methods for exposing previously emerged internal structures.

3. Testing Feature Creation Factors with an Input Activation Search

The first method we propose for discovering hidden structure is based on searching the network's input space for an optimal input pattern that will produce a designated target output. By analyzing the pattern of input activation required to produce a specific output, part of the system's inner connectivity might be revealed. We term this "mental" search method the *Input Activation Search.*

3.1. A Neural Network implementation of an Input Activation Search

The previously described neural network was used to test whether the Input Activation Search can reveal the pair-feature structure of the hidden layer. Input Activation Search was implemented by iteratively increasing the activation of the input units (p_i) that maximally decrease the difference between the ensuing activation values of the simulated output and the desired target output (t_i). Due to the nonlinearity of the internal units (e.g. h_2), we expect that once one of the connected input elements (p_4) is activated (due to random weight fluctuations), the other connected input element (p_3) will be statistically more effective than the remaining, unconnected input elements (p_1&p_2) in activating the target output. Thus, a paired input activation structure will emerge.

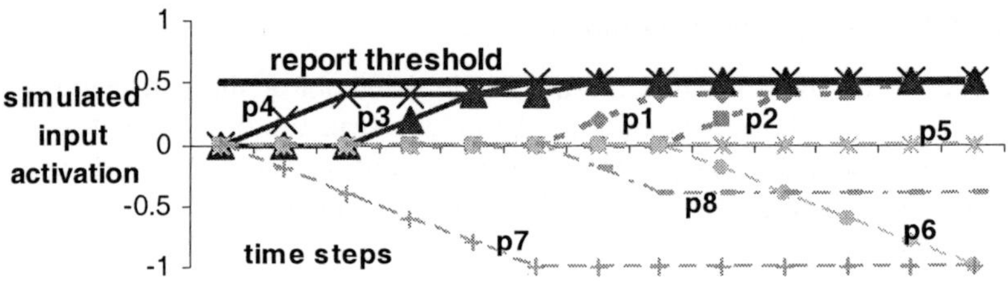

Figure 5: A simulation of the Input Activation Search revealing a congruent pair pattern in reporting category 1. Input elements were initialized at 0 ($p_{i,i=1,..,8}$=0). At every time step of the search a small increase of activation of each input unit is tested to see how well it minimizes the difference between the current output and the designated target output. Only the most potent increase is actually performed (p_i=p_i+ ε). Input activation of the units is bounded by an arbitrary value of ½ (simulating the reduced strength of imagery). When an input unit reaches this level it is regarded as if that input has been reported and therefore is no longer strengthened in the search process. A similar process of input deactivation (p_i=p_i-ε) simultaneously inhibits the non-relevant inputs.

We discovered that the input elements were consistently activated in a pattern congruent with the pair-features (while quadruple detector representations generated random input activation structures). We therefore conclude that Input Activation Search results can reflect the system's inner structure (Figure 5).

3.2. Experiment 1: Input Activation by Color Recall

The first experiment was designed to test whether our Input Activation Search method will detect a *pair-feature* structure. 27 psychology students with normal or corrected to normal vision completed the four learning stages described above in a 2-3 hour learning session. After the initial learning session participants were requested to verbally report the relevant color-cubes for each target button. We expect the sequence of color-cube reports to indicate whether a representation of the pair-features has emerged.

Figure 6: An example of the three patterns of reporting the first two cubes in category 1 (notice that the Adjacent & Congruent and the Adjacent & Incongruent pairs are spatially identical due to the perspective rotation employed in the experiment).

The first two color-cubes reported in each category can match one of three patterns (Figure 6). If a quadruple detector set is the sole representation emerging from the learning stage, we would expect an equal number of Adjacent & Congruent and Incongruent reports (Diagonal & Incongruent reports have a spatially different gestalt and cannot be compared to the other patterns). On the other hand, the relative frequency of Adjacent & Congruent reports compared to the spatially similar Adjacent & Incongruent reports indicates whether the *pair-features* have acquired a salient internal representation.

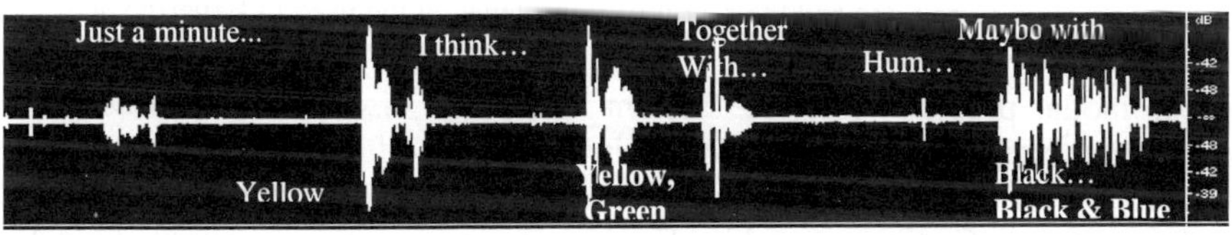

Figure 7: A 30-second voice recording of a category report complying with the congruent pair-feature pattern ("**Yellow-Green**" & "**Black-Blue**"). This pattern appeared in 90% of the participants' reports (notice the mental search periods that precede the *pair-features*).

Results: All participants succeeded in reporting the four colors relevant to each category. The frequency of reporting the congruent pattern was significantly higher ($p<0.01$, in a binomial test, $n=27$) than that of the spatially similar

Adjacent & Incongruent patterns in all four categories (Figure 7). It should be emphasized that the feature creation process could not have originated from any internal stimuli regularity like co-occurrence of color pairs, because these regularities were fully controlled for. In addition, in each trial the pictures were presented from a random point of view, thus canceling horizontal-vertical biases.

Although participants are explicitly required to verbally report each category, the sequence of reported color-cubes is an implicit measurement. We therefore believe that in addition to reflecting the internal structure that has emerged in the learning process, the reporting sequence was not intentionally or unintentionally biased by the subjects.

4. Testing Feature Creation by Additional Learning Facilitation

The Input Activation method demonstrated that an internal representation of pairs has been created, but it did not indicate whether this representation could be used as a tool in future category learning. We suggest that if learning future categories based on these pairs will be significantly facilitated, the internal representation of *pair-features* emerged as a functional tool in future learning. This method will be termed the Additional Learning Facilitation method.

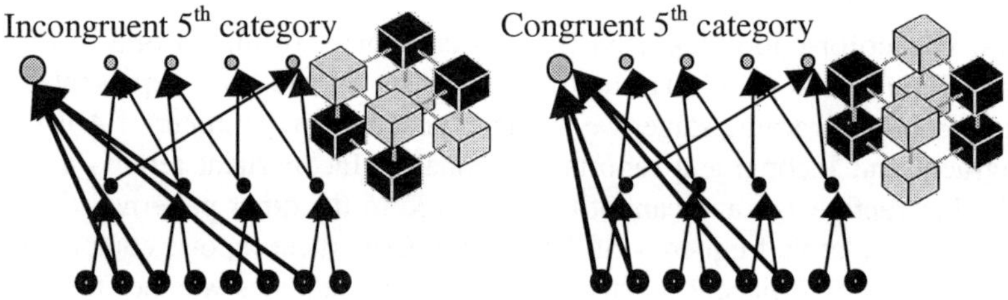

Figure 8: The input elements defining the 5th category in the congruent (right) & incongruent (left) conditions and their empirical implementation used in Experiment 2.

In our model problem, after the system learned to discriminate between the initial four categories we may require it to learn a fifth category either in a congruent condition or in an incongruent condition. In the congruent condition the fifth category $C_5 \equiv \{p_{i,i=1,2,5,6}=1\} \rightarrow \{t_5=1\}$ is composed of a new conjunction of two *pair-features* (pairs that had previously appeared in two learned categories). In the incongruent condition the fifth category $C_5 \equiv \{p_{i,i=2,3,6,7}=1\} \rightarrow \{t_5=1\}$ is composed of a conjunction of two other pairs each appearing in just one category (Figure 8). Unlike the first training session, the second training was limited in time, while monitoring the error rates of the systems in the congruent and incongruent conditions. If the congruent category is learned consistently faster than the incongruent category, we conclude that the pair-feature structure emerged as a functional tool in perceptual learning.

4.1. Neural Network implementation of Additional Learning Facilitation

The feasibility of the Additional Learning Facilitation method was tested using the neural network presented above, with one additional output. The fifth output unit was required to remain inactive (t_5=-1), while the first four categories were learned. Then, a fifth category was trained either in a congruent or in an incongruent condition. The second training was limited to three epochs (three presentations of the training set). In a set of 100 randomly initialized simulations, 65 of the networks in the congruent condition learned the new category, reducing the classification error to zero. None of the networks learning a fifth category in the incongruent condition managed to correctly classify the new category's exemplars [11].

4.2. Experiment 2: Additional Learning Facilitation

Experiment 2 was aimed at demonstrating that the *pair-features* can facilitate future category learning. After learning the initial four categories, 10 participants were randomly divided into two groups. Each group learned a fifth category based either on congruent *pair-features* $C_5 \equiv \{p_{i,i=1,2,5,6}=1\} \rightarrow \{t_5 = 1\}$ or on adjacent incongruent pairs $C_5 \equiv \{p_{i,i=2,3,6,7}=1\} \rightarrow \{ t_5 = 1\}$ (Figure 8). The stimuli presentation process was identical to that used in the initial learning stage. Both groups were required to learn the fifth category using a similar fixed set of 48 pictures. After the additional learning stage, participants were required to verbally report the color-cubes composing the new category. If only quadruple detectors have previously emerged we would expect that the learning rate of the fifth category would be equal in both groups. On the other hand, if *pair-features* were created, they could be used to facilitate future learning of the new congruent categories. It was therefore hypothesized that under limiting learning conditions (48 trials) only the congruent group will be able to learn the new category.

Results: The learning rate was significantly higher in the congruent than in the incongruent condition ($p<0.05$, Fisher Exact Probability Test, n=10). Whereas the congruent group reported an average of 2.2 out of 4 correct color-cubes, i.e. it learned most of the new category's characteristics, the incongruent group learned, on average, only 0.8 out of the four color cubes present in category 5.

5. Discussion

Our empirical study demonstrates that internal features, distinct from the categories on which the participants were trained, are being formed. We focused on contrasting the roles of information value and parsimony of a feature set as factors influencing feature creation. To this end we designed a simple *model problem* requiring four new categories to be learned, each based on a conjunction of four input elements. If discriminative information were the sole factor in the feature creation process, only the more informative quadruple features would emerge. Experimentally we found that participants' reporting-patterns corresponded to an internal structure based on pair-features. This shows that the information content of features is not the sole factor determining feature creation. We propose that parsimony is the other principle that plays a dominant role in the feature creation process. Our second experiment demonstrated that the pair-features might be revealed not only by biasing reporting patterns, but also by actively facilitating additional category learning. This result suggests that the perceptual system maintains parsimony not only as a resource optimization requirement, but also as a tool to assist future generalization, because parsimonious features are more likely to appear in future categories.

Feature creation processes could be viewed as a bias of the perceptual system towards future generalization. As demonstrated in Experiment 2 this bias is usually beneficial; nevertheless one could generate a setting where an emerging feature set might reveal itself by hindering incongruent future perceptual learning. This prediction will be tested in future experiments.

We claim that this study exhibited, for the first time, that the creation of new features is not necessarily a result of frequent co-occurrence or of gestalt biases (like spatial proximity or color similarity). Furthermore, it is not necessarily induced by direct feedback. The pair-features in our experiments were created due to both their high discriminative information on the learned categories and their parsimonious structure.

Acknowledgments

MF's research is supported by the Israel Foundations Trustees (grant 032.1478)

References

1. A. Archambault, C. O'donnell and P. Schyns. Blind to Object changes: when learning the same object at different levels of categorization modifies its perception. *Psychological Science*. **10(3)** (1999)
2. E. R. Guthrie. *The psychology of learning*. Harper & Row, NY, (1935)
3. A. Tversky. Features of similarity. *Psychological Review*. **84(4)**, 327 (1977)
4. P. G. Schyns, R. L. Goldstone and J. P. Thibaut. The development of features in object concepts. *Behavioral and Brain Sciences*. **21**, 1 (1998)

5. O. Rosenthal, S. Fusi and S. Hochstein. Forming classes by stimulus frequency: behavior and theory. *Proceedings of the National Academy of Sciences*, **98(7)**, 4265 (2001)
6. P. G. Schyns, and L. Rodet, Categorization creates functional features. *Journal of Experimental Psychology: Learning, Memory and Cognition*. **23**, 681 (1997)
7. S. Ullman, M. Vidal-Naquet and E. Sali, Visual features of intermediate complexity and their use in classification. *Nature Neuroscience*. **5**, 682 (2002)
8. M. Fink, *www.vision.caltech.edu/fink/InputDist.pdf* (2003)
9. F. D. Foresee and M. T. Hagan. Gauss Newton approximation to Bayesian learning. *Proceedings of the International Joint Conference on Neural Networks*. (1997)
10. M. Fink, *www.vision.caltech.edu/fink/NN.pdf* (2003)
11. M. Fink, *www.vision.caltech.edu/fink/ALF.pdf* (2003)

DOES THE ENERGY SPECTRUM FROM GABOR WAVELET FILTERING REPRESENT SUFFICIENT INFORMATION FOR NEURAL NETWORK RECOGNITION AND CLASSIFICATION TASKS?

MARTIAL MERMILLOD

Cognitive Science Department, Université of Liège, Liège, Belgium,
Email: mmermillod@ulg.ac.be

NATHALIE GUYADER AND ALAN CHAUVIN

Laboratoire de Psychologie et Neurocognition, CNRS, UMR 510, Laboratoire des Images et des Signaux, CNRS, ESA 5083, INPG, 27 Av Félix Viallet, Grenoble
Email :nguyader@lis.inpg.f r; alan.chauvin@laposte.net

Recent results from neurophysiological studies [11] suggest that energy spectra (i.e., the square of the amplitude spectrum) could be a suitable way to simulate, in a physiologically plausible manner, the spectral integration of sensory neurons. In this paper, we show for a high-level cognition task, the adequateness of the energy spectrum as an image descriptor for neural network computations. We used a simulation of cortical complex cell functions as a perceptual model which extracts image information. In a first simulation, we tested the energy spectrum descriptors with a back-propagation auto-encoder. In a second simulation, we tested the same descriptors with a standard back-propagation heteroassociator. The results show a reliable ability of these two types of neural networks to categorize and to generalize prior training to new exemplars based on the information provided by the energy spectrum of natural scene images.

1. Introduction

Despite their obvious complementarity, there have been relatively few attempts [5,6,12,20] to combine methods from perceptual cognition with high-level cognition tasks. Usually, scientists working on high-level cognition like recognition, memory or categorization, use a raw description of the external environment [2,17,19]. On the other hand, scientists working on low-level visual perception [7,10,15,16] are generally unconcerned with implications in high-level cognition tasks. However, a reliable simulation of perceptual processes could be of crucial importance for high-level cognition. First, to simulate the bottom-up constraints of the human visual system on visual cognition, which is not dealing directly with the raw code provided by natural images. Second, because of computational constraints, it is not possible to directly use the raw grey level code of an image: a simple low resolution grey level image (256*256 pixels) results in a vector of size 65536. To overcome that computational constraint, it was suggested to simulate the mammalian visual processes using a compressed representation of the image. Third, directly using the grey level of each pixel as an image descriptor cannot provide good results for high-level cognition tasks; in fact, two images belonging to the same semantic category (for example, two images of a city) can have very different grey level intensity distributions.

2. Statistical data compression

One of the methods widely applied to compress visual information consists of extracting the vectors that explain the maximum variance of the image intensity distribution with a Principal Component Analysis [1,24]. The Principal Component Analysis (PCA) finds a new orthogonal space for describing a pattern of data; each axis of this new space corresponds to the direction in which data have a maximum of variance. Hence, each point in the original space can be specified in this new space by the value of its projection along each axis. Moreover, by linear combination of the first eigenvectors, it is possible to create "category prototypes" and to perform visual categorization by comparing new exemplars to these prototypes [1]. On the basis of that method applied on a gender categorization task of human faces, Abdi et al. [1] reported a correct classification rate of 91.8 % when associated with a perceptron and 90.6 % when associated with a Radial Basis Function network. An approach very similar to PCA was proposed by Cottrell [4] and consisted of using the hidden layer of a back-propagation auto-associator in order to compress the input data. It results in a non-linear compression of the input layer so called "holons". That method is comparable to the use of a PCA, but at a non-linear level.

Despite good categorization performance, statistical categorization algorithms like PCA suffer from a lack of biological plausibility. Especially at a perceptual level, recent neurophysiological data [7;14;15;16;] do not support statistical computation directly on the raw code of the images and various kinds of pre-processing seem to occur in the early stages of visual perception.

3. Visual data compression with neurocomputational model

A more physiologically inspired method to compress visual information was to simulate the mammalian primary visual system; cells of primary visual cortex are sensitive to orientation and spatial frequency information. Poggio and Edelman [22] have proposed using a Radial Basis Function (RBF) network with Gaussian receptive fields handling different parts of the input layer (a two dimensional input array). The hidden layer nodes integrate different Gaussian receptive fields and compute an output function depending on the data extracted by the Gaussian filters. That model allows a reliable recognition rate of the same object across different points of view and it seems to be a good simulation of the biophysical mechanisms. More recently, RBF networks were applied to the categorization of visual images [3;8;9]. Edelman & Duvdevani-Bar [8] reported two basic properties of the unsupervised algorithm. First, objects seem to form clusters based on object identity (and not by pose or illumination). Second, clusters seem to be grouped by similarity between corresponding objects like vehicles, animals, forming semantic or super-ordinate "super-clusters". Those properties have led to a mean correct categorization rate around 80 % depending on the categorization method applied (i.e. Winner-Take-All or k-Nearest Neighbor). In every case, the RBF representation produced a better accuracy ratio than measurement-space representation in categorization tasks. In other words, RBF networks could lead to a fair approximation of the human categorization performance. Moreover, it is possible to approximate the neuro-physiology of the mammalian visual cortex by the mean of the receptive fields handling different property of the visual field.

However, Jones and Palmer [15,16] have shown that Gabor functions are better candidates to simulate the V1 receptive fields than Gaussian functions. Several papers have shown how Gabor wavelet decomposition can compress visual information. Wiskott [25,26] proposed applying Gabor filters under specific locations in the original image. In other words, several jets of Gabor wavelets are convolved with the original image in the spatial domain in order to extract different orientation and spatial frequency information under those locations. That method is economical but requires the experimenter to select different specific locations in the pictures for filtering. Another way consists of applying the bank of Gabor wavelets on a sliding window across the whole image [5,6]. That method allows encoding information for the entire image, however, that process requires a second step to compress that information (for example, using a PCA).

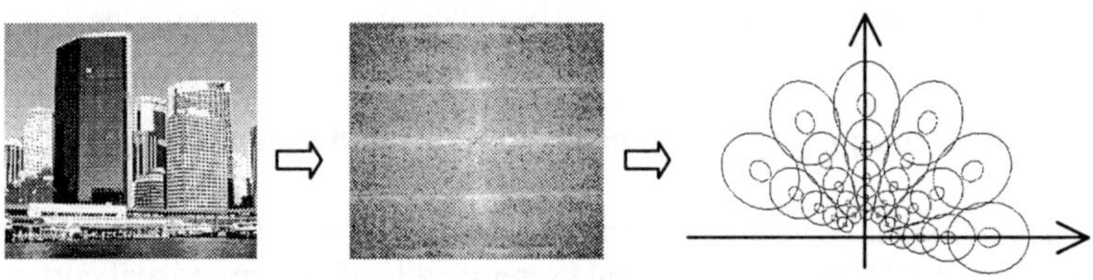

Figure 1: The perceptual model. Images are transferred in the Fourier domain and a bank of Gabor wavelets is applied in order to extract orientation and spatial frequency information.

4. The perceptual model

The method used in our model is to apply the bank of wavelets in the Fourier domain [14,21]. That method, which can be considered as a fovea vision simulation, allows the use of a single bank of Gabor wavelets for the whole image because the information coded in the Fourier domain is independent of spatial locations. Compared to other methods, this allows us not to decompose the original image (Figure 1).

The advantage of applying the Gabor wavelet filtering in the Fourier domain is a computational efficiency because there is no requirement of subsequent data compression after filtering. Moreover, as V1 complex cells, there is a spatial invariance of the Fourier domain. In other words, a Gabor wavelet applied in the Fourier domain is receptive for the orientation and spatial frequency information through the entire image. The result is very similar to the pooling process done by complex cell integration of simple cell responses. Simple cells are coding for specific orientation and spatial frequency under specific spatial locations of the visual field whereas complex cells are coding for the entire visual field. As complex cell responses, the Fourier domain is not sensitive to the spatial location of orientation and spatial frequency information. The descriptors used as outputs were the average energy value (i.e., the square of the amplitude spectrum) provided by each individual Gabor filter. However, a question remains about the validity of those descriptors for high-level cognitive tasks. In other words, do those descriptors contain sufficient information for categorization or recognition tasks? Gollisch & Herz [11] have shown that the energy spectrum is a reliable, neurophysiologically plausible way to simulate sensory neuron integration. At a computational level, recent studies [13,23] have shown that categorization tasks are possible using the entire power spectrum of natural images. In the present study, we are testing the reliability of the energy spectrum provided by the Gabor wavelet filtering, which can be considered as a compression in the Fourier domain, depending on the orientation and spatial frequency coded by each filter.

The stimuli used correspond to six categories of twelve natural scene images: Beach, City, Forest, Mountain, Indoor and Village. We first applied a Hamming window to avoid a boundary effect (i.e. an over representation of the vertical and horizontal orientations in the Fourier domain). After pre-processing, the images were transferred into the Fourier domain by the mean of a Fast Fourier Transform and filtered by a set of Gabor filters. Filter sizes were normalized with respect to a 1/f decreasing of the amplitude spectrum for natural images [10]; HF filters are wider than LF ones to compensate for 1/f decreasing. The filtered outputs were used to determine energy coefficients by coding the local energy spectra, which is the square of the amplitude spectra. We applied a single bank of thirty Gabor wavelets corresponding to five spatial frequency bands and six different orientations. According to the biological data about complex visual cells, the relative radial bandwidth of the Gabor filters was kept at one octave; the relative transversal bandwidth was 180°/6, in order to compute a bank of Gabor filters which covers the whole spectral domain but in accordance with neurophysiological data [7].

5. Connectionist simulations

5.1. *Back-propagation auto-encoder*

The first simulation was run on the auto-associator proposed by Mareschal and French [18]. That model, which could be considered as an unsupervised neural network, was proposed to simulate three-month-old infant categorization processes (a pure perceptual classifier, without access to any conceptual knowledge). The length-30 vectors were autoassociated by a 3-layer back-propagation network. The network architecture consists of 30 input units, 30 output units and 20 hidden units. The learning rate was fixed to 0.1, momentum to 0.9. The goal of the back-propagation network is to create an "internal representation" at the hidden layer level by the auto association of input and output vectors. We generated the internal representation by training the network on 11 exemplars of each training category. The training consisted of associating each exemplar with itself for 500 epochs. The neural network was trained on one category at a time, and then tested on the remaining exemplar from the training category versus one exemplar from the five other categories. Results were averaged over 100 runs of the above training-test procedure. The measure was the mean Euclidian distance between the outputs produced by the tested vector and the desired output. If new exemplars from the same category are correctly classified by the auto-encoder, we should observe a small increase in error between the observed output produced by the network and the desired output [18]. However, if exemplars from a novel category are recognized as different from the previously learned category, we should observe a significant increase of error.

In order to compute a classification rate (Figure 2), we applied a winner-takes-all algorithm on the Euclidean distances produced by the network (i.e., exemplar was classified in the category where it produced the lower Euclidean distance.)

The unsupervised algorithm produced a good rate of correct categorization for the Forest category exemplars (92 % of correct categorization), reasonable for Beach and Indoor (63 and 60 % of correct categorization) but weak for City, Village and Mountain (45; 39 and 34 % of correct categorization). As a recognition task, there was 100% correct recognition of a specific exemplar compared to the other exemplars from the same or other categories. The weak performances on the classification task could probably be improved with a segregation of the different parts of images at a perceptual level. For example, numerous mountains contain some forests on their slopes, which could result in a misclassification by the neural network. However, we decided to test a back-propagation heteroassociator, more suitable for classification tasks.

	NEURAL NETWORK CLASSIFICATION					
INPUT VECTOR	Forest	City	Indoor	Mountain	Village	Beach
Forest	**92**	0	0	7	0	1
City	0	**45**	24	0	31	0
Indoor	0	11	**60**	0	29	0
Mountain	39	0	0	**34**	8	19
Village	1	8	52	0	**39**	0
Beach	26	0	6	1	4	**63**

Figure 2: Classification rate produced by the autoassociator.

5.2. *Back-propagation heteroassociator*

This new simulation was run with a classic 3-layer back-propagation heteroassociator. The network architecture consisted of 30 input units, 20 hidden units and 6 output units, each output node coding for one of the 6 training category. As in the previous simulation, the learning rate was fixed to 0.1, momentum to 0.9. The goal of the heteroassociator is to associate each category exemplar to the output vector coding for that specific category. Thus, the network was trained on all 66 exemplars, keeping one remaining exemplar per category for the generalization test. The network was trained for 50 epochs (convergence occurring faster with heteroassociation) and results were averaged over 100 runs. The measure was the output produced by each individual output nodes and a winner-takes-all algorithm was applied to the output values in order to compute a classification rate (Figure 3).

The classification rate produced by the heteroassociator was a hundred percent correct classification for City, Mountain and Beach exemplars, 98 % of correct classification for Indoor and 86 % for Forest. These results suggest a sufficient reliability of the input vector composed of the energy information provided by the Gabor filters for distributed categorization processes. As in the previous simulation, those results could be improved with the use of preprocessing, teasing apart the different semantic elements constituting an image.

	NEURAL NETWORK CLASSIFICATION					
INPUT VECTOR	Forest	City	Indoor	Mountain	Village	Beach
Forest	**86**	0	0	12	2	0
City	0	**100**	0	0	0	0
Indoor	0	0	**98**	0	2	0
Mountain	0	0	0	**100**	0	0
Village	0	0	10	0	**90**	0
Beach	0	0	0	0	0	**100**

Figure 3: Classification rate produced by the heteroassociator.

6. Conclusion

Among the different techniques used to compress visual information, Gabor wavelet filtering seems to be one of the most biological plausible ways to reproduce the mammalian visual pathway [15,16]. Running the filtering in the Fourier domain allowed us to use a single bank of Gabor filters to encode the entire image. However, the descriptor extracted by that perceptual model (i.e. the average energy value provided by each filter) had to be tested in a high level classification and recognition task in order to show that those descriptors contained sufficient information for subsequent distributed processes. In our initial connectionist simulation, we showed that the energy vectors are sufficiently discriminable to be perfectly differentiated with a standard back-propagation autoassociator. On the other hand, the autoassociation process did not produce the best performance in a categorization task. In a second simulation, a classic backpropagation heteroassociator allowed us to obtain reliable accuracy in such categorization tasks. Obviously, the heteroassociation seems to be better in categorization tasks, despite an ecological advantage (because of its unsupervised learning procedure) of the autoassociator to simulate early visual categorization. This suggests that the exemplars of the same category are sufficiently similar at the level of their energy spectrum to allow a distributed network to learn those perceptual structures in order to perform a reliable categorization performance on new exemplars coming from the same or different categories.

7. Acknowledgments

This work has been supported by European Commission grant HPRN-CT-1999-00065. Thanks to Robert French for helpful comments.

References

1. Abdi, H., Valentin, D., Edelman, B.E., O'Toole, A.J. (1995). More about the difference between men and women: Evidence from linear neural networks and the principal component approach. *Perception*, 24, 539-562.
2. Anderson, J. R., & Fincham, J.M. (1996). Categorization and sensitivity to correlation. *Journal of experimental psychology: Learning, Memory and Cognition, 22,* 259-277.
3. Basri, R., Roth, D., & Jacobs, D (1998). Clustering appearances of 2d objects. *In IEEE Conference on Computer Vision and Pattern Recognition, Santa Barbara, CA,* 414-420.
4. Cottrell, G.W. (1990). Extracting features from faces using compression networks: Face, identity, emotion, and gender recognition using holons. *In Proceedings of the 1990 Connectionist Models Summer School,* (eds. D. Touretsky, J. Elman, T. Sejnowski & G. Hinton) *Kaufman,* 328-337.
5. Cottrell, G. W., Branson, K. & Calder A. J. (2002). Do expression and identity need separate representations? In Proceedings of the 24th Annual Cognitive Science Conference, Fairfax, Virginia. Mahwah: LEA
6. Dailey, M. & Cottrell G. (1999). Organization of face and object recognition in modular neural network models. *Neural Networks, 12,* 1053–1073.
7. De Valois, R.L., & De Valois K.K. (1988). *Spatial Vision.* New York: OUP.
8. Duvdevani-Bar, S., & Edelman S. (1999). Visual recognition and categorization on the basis of similarities to multiple class prototypes. *International Journal of Computer Vision, 33(3),* 201-228.
9. Edelman, S. & Duvdevani-Bar, S. (1997). A model of visual recognition and categorization. *Proceedings of Royal Society, London, vol. B-352,* 1191-1202.
10. Field, D.J. & Brady N. (1997). Visual sensitivity, blur and the sources of variability in the amplitude spectra of natural scenes. *Vision Research, 37.* 3367-3383
11. Gollisch T., & Herz A.V.M. (2003). Analyzing mechanosensory transduction by identifying invariant directions in stimulus space. *Neurocomputing, 52(54),* 525-53.
12. Gosselin, F. & Schyns P.G. (2002). R.A.P.: a new framework for visual categorization. Trends in Cognitive Science, 6(2), 70-77.

13. Guérin-Dugué, A., Oliva, A. (2003). Classification of scene photographs from local orientations features. *Pattern Recognition Letters, 21,* 1135-1140.
14. Guyader, N. Hérault J. (2001). Représentation espace-fréquence pour la catégorisation d'images. In Proceedings of the 18th GRETSI Symposium on Signal and Image Processing, Toulouse, France.
15. Jones, J.P. & Palmer L.A. (1987). The two-dimensional spatial structure of simple receptive fields in cat striate cortex. *Journal of Neurophysiology, 58,* 1187-1211.
16. Jones, J.P., Stepnoski A. & Palmer L.A. (1987). The two-dimensional spectral structure of simple receptive fields in cat striate cortex. *Journal of Neurophysiology, 58(6),* 1212-1232.
17. Kruschke J.K. (1992). ALCOVE: an exemplar-based connectionist model of category learning. *Psychological review, 99*, 22-44.
18. Mareschal, D. & French, R. (1997). A connectionist account of interference effects in early infant memory and categorization. *Proceedings of the 19th Annual Cognitive Science Society Conference*, LEA, 484-489.
19. McClelland, J. L., McNaughton, B. L. & O'Reilly, R. C. (1995). Why there are complementary learning systems in the hippocampus and neocortex: Insights from the successes and failures of connectionist models of learning and memory. *Psychological Review, 102(3)*, 419-437.
20. Morrison D.J. & Schyns P.G. (2001). Usage of spatial scales for the categorization of faces, objects and scenes. *Psychonomic Bulletin & Review, 8,* 454-469.
21. Oliva, A. & Schyns P.G. (1997). Coarse blobs or fine edges? Evidence that information diagnosticity changes the perception of complex visual stimuli. *Cognitive Psychology, 34,* 72–107.
22. Poggio, T. & Edelman, S. (1991). A network that learns to recognize three-dimensional objects. *Nature, 343*, 263-266.
23. Torralba, A. & Oliva, A. (2003). Statistics of natural image categories. *Network: Comput. Neural Syst., 14,* 391–412.
24. Turk, M. & Pentland, A. (1991). Eigenfaces for recognition. *Journal of Cognitive Neuroscience, 3(1),* 71-86.
25. Wiskott L. (1997). Phantom Faces for Face Analysis. *Pattern Recognition 30(6),* 837-846.
26. Wiskott, L., Fellous, J.M., Krüger, N. & Von der Malsburg C. (1999). Face Recognition by Elastic Bunch Graph Matching. *In Intelligent Biometric Techniques in Fingerprint and Face Recognition, eds. L.C. Jain et al., CRC Press, 11,* 355-396.

Attention

THROUGH ATTENTION TO CONSCIOUSNESS BY CODAM

JOHN G. TAYLOR

Department of Mathematics, King's College, University of London

Strand, London, WC2RS2LSe, UK

E-mail: john.g.taylor@kcl.ac.uk

Consciousness is best approached through attention. An engineering control approach to attention and motor response is presented here, culminating in the CODAM model of consciousness. This is briefly described, as is support arising from brain dynamics, especially that for the attentional blink.

1. Introduction

Attention is known to be a control system par excellence: it singles out inputs to be processed to a higher level in a given modality. Attention involves two types of area in the brain: those being controlled, and those achieving the control. The control is by amplification of required inputs with inhibition of distracters, now considered achieved by competitive feedback from controlling to controlled areas, with modulation by overall multiplication of controlled neural activity, or by biasing response threshold.

It was suggested [18] that engineering control provides a framework to understand attention, as will be reviewed in the next section. Simulations have been performed in this approach, both for sensory attention [23] and extended to sensory-motor attention, as described briefly in section 3.

The engineering control approach to attention has been developed into a model for the formation of conscious experience, by means of the COrollary Discharge of Attention Movement (CODAM) [18, 19, 20, 21, 22]. This approach is reviewed briefly in section 4, with experimental support briefly discussed in the following section, both in terms of timing in the brain and an explanation of the attentional blink. The final section contains further explanatory remarks about CODAM.

2. The Control Approach for Attention

There is now good experimental evidence that attention acts in a control manner, as noted in the introduction, with several distinct brain regions being involved. The attention-controlled region consists at least of sensory and motor cortices. The control module generating the necessary attention movement control signal, and denoted IMC (for inverse model controller), is sited in the parietal lobes.

Further Goal modules are needed to process external signals that draw attention, and also to hold internal goals; these are observed in the prefrontal cortex. A Monitor module to check that error level in directing attention is low, and which helps learn motor responses, is detected in the cingulate cortex. An Observer module, estimating the state of the controlled system (in lower-level cortex), is also likely to be present. Activity in this observer is based on the corollary discharge of the attention movement control signal (denoted WMcd, in sites yet to be determined). Finally a Feedback Processor is present, acting as a buffer, and holding the signal from the amplified sensory input for further processing such as report (and denoted WM Sensory); it is sited in the parietal lobes.

A standard engineering control circuit was developed for attention [18, 19, 22] by analogy to models of motor control in the brain [3, 14, 25]. Exogenous attention control is achieved through rapid passage of the input signal to the goal module, leading to feed-forward control from the IMC. Endogenous control arises by holding a suitable goal signal on the Goal module, enabling biasing of the competitive process on the IMC to take place to achieve attention as focused on the target input (such as detecting a face in a crowd).

The observer (WMcd) and the feedback processor (WM Sensory) are taken as working memory sites [1]. These two modules are specializations to the brain of the more general observer and feedback systems well known in engineering control. There are a variety of ways in which these observer and feedback modules could combine their outputs to the IMC, as well as its combination with more direct feed-forward control signals directly from the Goal module to the IMC.

Considerable experimental support exists for most of the attention control modules introduced above. This comes from numerous brain imaging and single cell experiments [2, 6, 8, 9], to which the reader is referred.

3. Sensory-Motor Attention Processing

It has been suggested by Rushworth [4, 12] that there exists a distinct mechanism of motor attention localized in the anterior parietal cortex. This mechanism involves pre-motor areas that control limb movements, and parietal areas, such as the supramarginal gyrus (SMG), where effector-centered, rather than head-centered, visuospatial representations can form. Rushworth and colleagues found motor attention-related activity not only localized anterior to the area concerned with orienting, but also laterally in the left hemisphere.

An extension of a simplified version of the above control system to include motor attention is shown in figure 1 [24]; for simplicity the new control system has been shown without the buffer observer modules described above.

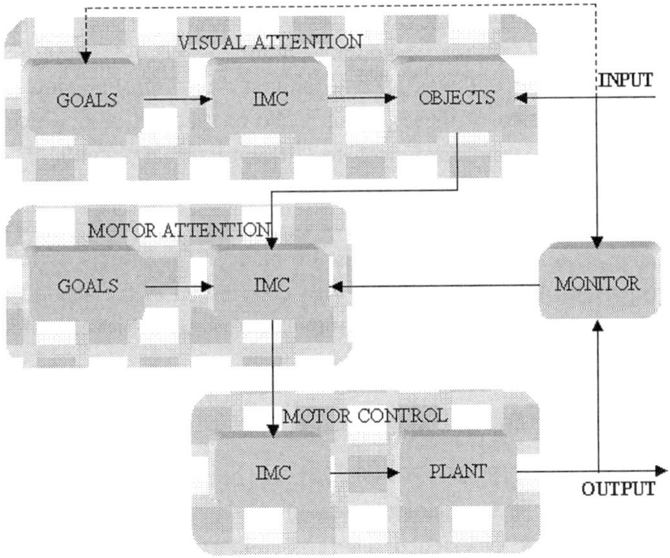

Fig 1. A model of sensory-motor attention control. The notation is in the text.

In figure 1 we assume that in the pathway from the visual input-to-motor response there are three stages of processing, which involve visual attention (deciding where/ what to attend to), motor attention (specifying where/what to act on) and automatic motor control (implementing the attended motor action).

The model is an extension of the sensory attention control framework of section II to motor attention. The modules present in figure 1 are: a) the inverse motor attention model controller (IMC) module, extending that for sensory attention to its motor companion. This module receives the object data selected and amplified in the visual attention processing stage and combines it with motor response, in order to guide the motor control IMC. It is biased by the associated motor attention goals module (distinct from the visual attention goals module) and monitored by the monitor module so as to learn the appropriate visuo-motor responses; b) The motor attention goals module, which contains relevant visuo-motor combinations according to the rules of the task and is used to bias their learning within the IMC; c) The monitor module, which checks for errors by comparing the motor output with the visual input, and then sends a training signal to the IMC which increases the learning when the error rate rises and decreases the learning when the error rate falls; d) The inverse motor model controller (IMC) module, driven by the motor attention IMC via control signals that correspond to the motor response selected in the motor attention processing stage, which is translated into a motor command that controls the musculature; e)·The controlled motor cortex/spinal chord motor signals of the motor system.

The model of figure 1 was used to simulate two sensory-motor attention paradigms [24]. The first of these [13] is a choice reaction task, which requires subjects to make different responses to different visual stimuli. The second paradigm [15] concentrated on a motor preparation task, determining the benefit in reaction time gained by valid motor responses compared to invalid ones (where change to a valid response must be made). The effects of left or right hemisphere parietal lobe deficits were also determined in the second paradigm from patients with relevant deficits. Both simulations achieved reaction time values close to experimental values.

4. The CODAM Model of Consciousness

It has been proposed [18, 19, 20, 21, 22] that the corollary discharge buffer WMcd of section 2 is a crucial element in the circuitry of a model for the formation of conscious experience. There is support for the presence of such a corollary discharge component in the overall attention control circuitry by analogy with motor control, from the evidence for predictor modules for such control 3, 14, 25]. We therefore introduce the WMcd module as a crucial component to give the sense of ownership in conscious experience, so the module is a crucial pillar in the CODAM model.

We require there to be an inhibitory process between the observer, WMcd, and the sensory buffer, WM sensory. Further detail has to be given as to how processing in the observer/buffer/monitor sites can support conscious experience. One possibility has been outlined in [18, 19, 20, 21, 22]: the initial corollary discharge signal on WMcd, when arriving there, creates an expectation of future input from the relevant sensory cortex, and provides support to the signal from lower cortex trying to access the sensory buffer. The latter is amplified over some 100 msec or more [9], so as to be able to overcome inhibition from previous input activations in various sites, especially the WM Sensory buffer itself. When the amplified cortical signal becomes large enough, it can overcome the barrier to its entry to WM Sensory, so that awareness of the input is then supposed to arise. The activity on WM Sensory then also inhibits WMcd, so as to be able to be used for the next target. However before that occurs, awareness of the attention movement itself has arisen, as the experience (by the signal on WMcd) of ownership of the about-to-arrive content. The gap between the pre-reflective self of Western phenomenology and Eastern Buddhism and the content-full consciousness of Western cognitive neuroscience can be bridged in this manner, ass explored more fully in [19, 20].

5. Evidence for CODAM: Temporal Flow

The crucial component of the architecture needed for CODAM, beyond the modules already known to be present (sensory, IMC, goal, monitor and WM Sensory modules), is the buffer site WMcd. There are already preliminary hints for the existence of an early cortical signal reviewed in [6], as well as new data presented there. A brief overview will be given here of the temporal flow of activity in the brain, based on EEG, fMRI and MEG data [5, 8, 16, 17].

A sequence of peaks and troughs are observed post-stimulus in averaged EEG brain signals: P1(80-120ms)/ N1(140-180ms), N2, N4(300ms), P3(300-500ms). The P1/N1 complex is attenuated for extinguished as compared to consciously observed contralesional stimuli on bilateral trials. Also in the attentional blink (AB) (in which there is rapid input of the second of two targets, T2, during the 'blink' period when the first target, T1, is being processed) there is unattenuated P1/N1 or N400 for T2, for various T1/T2 asynchronies, although the P300 was strongly reduced at the blink, as was the N2 signal [8, 16, 17].

The conclusion on ERP functionality, from a variety of paradigms in healthy individuals, is as follows:

- P1/N1 involves initial overall brain processing (with various sources of the P1/N1, both posterior and anterior).
- N2 corresponds to attention processing.
- N4 arises from semantic spreading.
- P3 is due to access to buffer working memory sites for higher processing (as the signal of awareness).

A recent MEG study [5], performed during a conjunction search paradigm, has shown important structure and topography in the N2 signal. It was found to consist of at least two components: an early parietal source (180-200 ms) and a later occipito-temporal source (220-240 ms). The first is consistent with activation of IMC to move attention; the second is involved with object/feature analysis. The crucial CODAM signal is that of the WMcd refreshment during movement of the focus of attention, and may already have been seen in this N2 signal, although the site of the WMcd access signal is unclear.

The CODAM model has been applied to the attentional blink (AB) paradigm [8, 16, 17]. The AB arises in the case of rapid serial visual presentation, with about 90-00 msecs gap between stimuli. There is a maximum lose of the ability to detect a second target, T2, in the stimulus stream (such as a white letter), about 250 msecs after the correct initial detection of a first, T1, such as a white X [16]. The N2 signal is lost in the AB, as is the P3, while the N1 and the N400 are preserved.

We note that the existence of the AB depends on the presence of a mask M1 for T1 & M2 for T2. Thus if there is no M1, then there is a much-reduced AB, while there is no AB if there is no M2. This implies that the 'scarce resource' character of attention is emphasized through masking, when M1, in particular, causes severe distraction to T1 detection, leading to the AB. We therefore need to protect T1 against M1 damage, with resulting greater inhibition of T2. Such a mechanism arises naturally in CODAM if the monitor is used to feed back an early error estimate, equal to the difference of the goal activation level and that of the WMcd, so providing an extra resource to the incoming input T1 to attain its sensory buffer. The resulting simulation [4] leads to good agreement with AB data in association with masking.

In more detail, the goals for T1, T2 are separated into two distinct nodes for each target: one for the exogenous bias of attention (identical to the masks' goal node), and one for the endogenous bias of attention. The latter is fed a constant pulse with amplitude of 5% of that of the input pulse (which drives the exogenous goals), and starts before any stimulus onset, lasting for about 2secs. The two nodes for the endogenous bias of attention for T1 and T2 are not allowed to be active at the same time, by requiring the endogenous T1 goal to inhibit the endogenous T2 goal until the endogenous T1 goal is turned off by a signal that arises from the T1 WM sensory buffer activation crossing a suitable threshold. This occurs at about 500ms after T1 stimulus onset.

The output of the endogenous and the exogenous goals for each of the two targets is added together jointly to bias the IMC through the monitor module. The monitor compares the endogenous goal activations for the two targets with their corresponding corollary discharge activations. It captures the endogenous goal activations only at the interval that the exogenous goals are active and buffers them to compare with the corollary discharge activations that follow. The difference of the two signals - the error signal - is used to compensate for damage from distracters, and especially to protect the targets T1 and T2 from further damage from the following incoming stimuli by being fed into a self-recurrent neuron. That spreads inhibition onto all the nodes in the IMC, apart from the target node under protection. In the case of T1, in particular, the nodes that will be inhibited by this protector neuron include the secondary target T2, which at that time poses a potential threat to T1's further processing. This inhibition lasts until confirmation arrives from the WM sensory buffer that the target has reached a significant level of activation, i.e. awareness.

The discrepancy for the T1 monitor node is constant across the lags because it only arises from the deterioration of the IMC signal caused by the battle (in the IMC) with the mask M1. On the other hand, the discrepancy for the T2 monitor node varies across the lags and depends on the level of (endogenous) goal activation that T2 has been allowed to have (by the mechanism described above) and the level of corollary discharge activation that survives the inhibition caused by the WM sensory buffer activation of T1. For the parameters chosen for the simulation, the difference between the two signals is almost zero for those lags and doesn't elicit an error signal in the monitor, so not helping T2 as it would need to compete with T1. This is the cause of the attentional blink in CODAM, and more specifically its dependence on the presence of the mask M1.

6. Conclusions

The use of an engineering control approach to attention was shown to lead to a general control model of the movement of attention whose simulation explained several attention paradigms, both for visual and visuo-motor attention movement. The control framework was extended to a buffer for an efferent copy of the attention movement control signal; this signal was interpreted as supporting the experience of ownership of the about-to-arrive signal of the content of awareness. The model helped explain how the attentional blink, a very attention-demanding task, is sensitive to masking of the first target.

Various features of this model need further detailed analysis to support its character as a general model of attention control, and more crucially to give more support for CODAM as a model of the formation of conscious experience. This has elsewhere been explored from the viewpoint of inner experience of schizophrenics in the early stage of their disease [21]. The loss of a sense of an inner self, in particular, needs to be explored by means of brain measurements on such patients, so as to relate to those regions of the brain that CODAM would suggest would be damaged or under-developed. The most crucial of these is clearly the WMcd component.

How does CODAM begin to catch the essential elements of consciousness? Attention is the gateway to consciousness, and this is incorporated in CODAM, which is an extension of the basic attention control model of figure 1 by inserting suitable working memory buffers, as noted in section 2. One of these, the sensory buffer WM Sensory, has been suggested by many as being the essential site for the formation of conscious experience, and is related to the P300, regarded as a signal of that experience. However that only gives a holding site for temporally extended activity, acting as content of consciousness. The crux of CODAM is the further signal, interpreted as the earlier N2/P2 signal, which grants the owner the experience of ownership of the about-to-arrive contents. This early signal grants the crucial property of 'Immunity to error due to misidentification of the first person pronoun', arguably being, together with attention, sufficient attributes for consciousness. This immunity is granted by the detailed interactivity of the corollary discharge of the movement control signal with the incoming sensory content, as seen from the discussion in section 4.

References

1. Baddeley, A. (1986). Working memory. Oxford: OUP.
2. Corbetta, M., & Shulman, G. L. (2002). Control of goal-directed and stimulus-driven attention in the brain. *Nature Reviews Neuroscience, 3*, 201-215.
3. Desmurget, M., & Grafton, S. (2000). Forward modelling allows feedback control for fast reaching movements. *Trends in Cognitive Neurosciences, 4*, 423-431.
4. Fragopanagos, N, Kockelhoren, S. & Taylor J. G. (in preparation) Simulating the dynamics of the attentional blink.
5. Hopf, M., Luck, S., Girelli, M., Hagner, T., Mangun, G., Scheich, H., & Heinze, H.J. (2000). Neural sources of focused attention in visual search. *Cerebral Cortex, 10*, 1233-1241.
6. Ioannides, A. A., & Taylor, J. G. (2003). Testing models of attention with MEG. *Proceedings of IJCNN'03.*
7. Kastner, S., & Ungerleider, L. G. (2000). Mechanisms of visual attention. *Annual Reviews of Neuroscience, 23*, 315-341.
8. Luck, S. J., Woodman, G. F., & Vogel, E. K. (2000). Event-related potential studies of attention. *Trends in Cognitive Sciences, 4*, 432-440.
9. Mehta, A. D., Ulbert, I. & Schroeder, C. E. (2000). Intermodal selective attention in monkeys II: Physiological mechanisms of modulation. *Cerebral Cortex, 10*, 359-370.
10. Nobre, A. C. (2001). The attentive homunculus: now you see it, now you don't. *Neuroscience and Biobehavioral Reviews, 25*, 477-496.
11. Rushworth, M., Ellison, A., & Walsh, V. (2001). Complementary localization and lateralization of orienting and motor attention. *Nature Neuroscience, 4*, 656-661.
12. Rushworth, M., Krams, M., Passingham, R. (2001). The attentional role of the left parietal cortex: the distinct lateralization and localization of motor attention in the human brain. *Journal of Cognitive Neuroscience, 13*, 698-710.
13. Rushworth, M., Nixon, P., Renowden, S., Wade, D., & Passingham, R. (1997). The left parietal cortex and motor attention. *Neuropsychologia, 35*, 1261-1273.
14. Sabes, M. (2000) The planning and control of reaching movements. *Current Opinion in Neurobiology, 10*, 740-746.
15. Schluter, N., Krams, M., Rushworth, M., & Passingham, R. (2001). Cerebral dominance for action in the human brain: The selection of actions. *Neuropsychologia, 39*, 105-113.

16. Shapiro, K. L., Arnell, K. M., & Raymond, J. E. (1997). The attentional blink. *Trends in Cognitive Sciences, 1*, 291-295.
17. Shapiro, K. L., Hillstrom, A. P., & Husain, M. (2002). Control of visuotemporal attention by inferior parietal and superior temporal cortex. *Current Biology, 12*, 1320-1325.
18. Taylor, J. G. (2000). Attentional movement: the control basis for consciousness. *Society for Neuroscience Abstracts, 26*, 2231#839.3.
19. Taylor, J. G. (2002a). Paying attention to consciousness. *Trends in Cognitive Sciences, 6*, 206-210.
20. Taylor, J. G. (2002b). From matter to mind. *Journal of Consciousness Studies*, 6, 3-22.
21. Taylor, J. G. (2003). Consciousness, neural models of. In M. Arbib (Ed.), *The Handbook of Brain Theory and Neural Networks*, pp. 263-267, Cambridge MA: MIT Press.
22. Taylor, J. G. (in press). Paying attention to consciousness. *Progress in Neurobiology*.
23. Taylor, J. G., & Rogers, M. (2002). A control model of the movement of attention. *Neural Networks, 15*, 309-326.
24. Taylor, J. G., & Fragopanagos, N. (2003). Simulation of attention control models of sensory and motor paradigms. *Proceedings of the IJCNN'03*.
25. Wolpert, D. M., & Ghahramani, Z. (2000). Computational principles of movement neuroscience. *Nature Neuroscience, 3*, 1212-1217.

MODELING VISUAL SEARCH: EVOLVING THE SELECTIVE ATTENTION FOR IDENTIFICATION MODEL (SAIM)

DIETMAR HEINKE, GLYN W. HUMPHREYS, CLAIRE L. TWEED*

Behavioural Brain Sciences Centre, School of Psychology, University of Birmingham, Birmingham B15 2TT, United Kingdom,
E-mail: d.g.heinke@bham.ac.uk

We present an extension of the Selective Attention for Identification model (SAIM) [1] in which feature extraction processes are incorporated. We show that the new version successfully models experimental results from visual search. We also predict the influence of a target cue on search. This extended version of SAIM may provide a powerful framework for understanding human visual attention

1. Introduction

Recently, we have presented a connectionist model of human visual attention, termed SAIM (Selective Attention for Identification Model) [1]. SAIM's behaviour is dominated by interactions between processing units within and between modules compete to control access to stored representations for object recognition. SAIM provides a qualitative account of a range of psychological phenomena on both normal and disordered attention. Simulations on normal attention match psychological data on: two-object costs on selection, effects of object familiarity on selection, global precedence, spatial cueing both within and between objects, and inhibition of return. When simulated lesions were conducted, SAIM also demonstrated both unilateral neglect and spatial extinction, depending on the type and extent of the lesion. Different lesions also produced view-centred and object-centred neglect, capturing the finding that both forms of neglect can occur within a single patient. In essence, SAIM suggested that attentional effects in human behaviour result from competitive interactions in visual selection for object recognition, whilst neurological disorders of

*This paper was supported by grants from the BBSRC, the EPSRC and MRC, UK.

selection are due to imbalanced competition following damage to areas of the brain modulating access to stored knowledge.

In this paper we present an extended version of SAIM in which a feature extraction process was added whilst at the same time maintaining the basic principles of SAIM, (e.g. competitive interactions and selection mechanism). Our aim here is to demonstrate that this new version still successfully performs translation-invariant object identification. Additionally, we assess the viability of 'extended SAIM' as a psychological model, testing whether it can simulate and explain data from human visual search tasks.

Visual search is a commonly-used paradigm in psychological studies of attention in which participants are asked to report the absence or presence of a specified target item amongst irrelevant items (distractors). Typically, performance is measured in terms of time until response (reaction time). The number of distractors is varied across trials. A typical outcome of many experiments is a linear function between reaction time and number of distractors. The slope of this linear relation is often interpreted as an indicator of the underlying search mechanism. For instance, a small slope (0-10ms/item) is interpreted as parallel search and a steep slope (20-50ms/item) is assumed to indicate serial search, based on one item at a time (see [2], for a recent review).

2. SAIM

2.1. *Overview*

Figure 1 gives an overview of SAIM's architecture and highlights the modular structure of SAIM with each module.

In the first stage features are extracted from the input image (the feature extraction process). The contents network maps a section of the features into a smaller Focus of Attention (FOA), a process modulated by spatial attention. In addition the mapping of the contents network into the FOA is translation-invariant, enabling SAIM to perform translation-invariant object recognition. The selection network controls the contents network by competitive interactions between its processing units, so that input from only one (set of) locations is dominant and mapped into the FOA. At the top end of the model, the knowledge network identifies the contents of the FOA using template matching. The knowledge network also modulates the behaviour of the selection process with top-down activation, so that known objects are preferred over unknown objects. In addition to these modules,

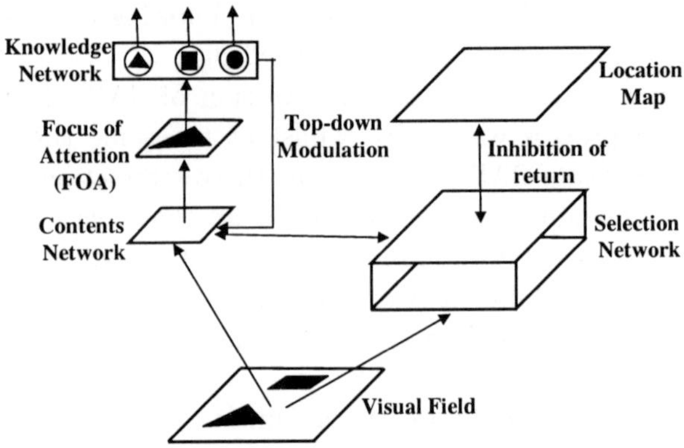

Figure 1. Architecture of SAIM

there is also a location map that enables SAIM to make multiple selections. Essentially units in the location map store the object position each time an object is recognized and then inhibits the selection network from reselecting these locations (inhibition of return).
The design of SAIM's network follows the idea of soft constraint satisfaction in neural networks based on "energy minimization" [3]. In SAIM the "energy minimization" approach is applied in the following way: Each module in SAIM performs a pre-defined task (e.g. the knowledge network has to identify the object in the FOA). In turn each task describes allowed states of activation in the network. These states then define the minima in an energy function. To ensure that the model as a whole satisfies each constraint, set by each network, the energy functions of each network are added together to form a global energy function for the whole system. The minima in the energy function is found via gradient descent, as proposed by [3]. In the following sections the energy functions for each network are stated. The global energy function and the gradient descent mechanism are omitted, since they are clearly defined by the subcomponents of the energy function.

2.2. *Feature extraction*

The feature extraction results in a three-dimensional feature vector: horizontal and vertical lines and the image itself. The lines are detected by filtering the image with 3x3 filters ($\begin{matrix} -2 & +1 & -2 \\ -2 & +1 & -2 \\ -2 & +1 & -2 \end{matrix}$ for vertical lines and its transposed version for horizontal lines). The feature vector is noted as f_{ij}^n hereafter, with indices i and j refereing to retinal locations and n to the feature dimension. This feature extraction process provides an approximation of simple cell responses in V1. As becomes obvious in the following sections, the use of just this simple feature extraction is not of theoretical value and arises only from practical consideration (e.g., the duration of any simulations). In principle, a more biologically realistic feature extraction process can be substituted (e.g. Gabor filter).

2.3. *Contents network*

The energy function for the contents network is:

$$E^{CN}(\mathbf{y^{SN}}, \mathbf{y^{CN}}) = \sum_{ijlm} (y_{lmn}^{CN} - f_{ij}^n)^2 \cdot y_{lmij}^{SN} \tag{1}$$

y_{lmij}^{SN} is the activation of units in the selection network and y_{lmn}^{CN} is the activation of units in the contents network. Here and in all the following equations the indices i and j refer to retinal locations and the indices l and m refer to locations in the FOA. The term $(y_{lmn}^{CN} - f_{ij}^n)^2$ ensures that the units in the contents network match the feature values in the input image. The term y_{lmij}^{SN} ensures that the contents of the FOA only reflect the region selected by the selection network ($y_{lmij}^{SN} = 1$). Additionally, since setting an arbitrary choice of y_{lmij}^{SN}s to 1 allows any location to be routed from the feature level to the FOA level, the contents network enables a translation-invariant mapping.

2.4. *Selection network*

The mapping from the retina to the FOA is mediated by the selection network. In order to achieve successful object identification, the selection network has to fulfill certain constraints when it modulates the mapping process. These constraints are that: (i) units in the FOA should receive the activity from only one retinal unit; (ii) activity of retinal units should be mapped only once into the FOA; (iii) neighbourhood relations in the retinal input should be preserved in mapping through to the FOA. Now, to incorporate the first constraint, that units in the FOA should receive the activity of only one retinal unit, the equation of the WTA-equation suggested by [4] turns into:

$$E_{WTA}^{SN1}(\mathbf{y^{SN}}) = \sum_{ij} (\sum_{lm} y_{lmij}^{SN} - 1)^2 \tag{2}$$

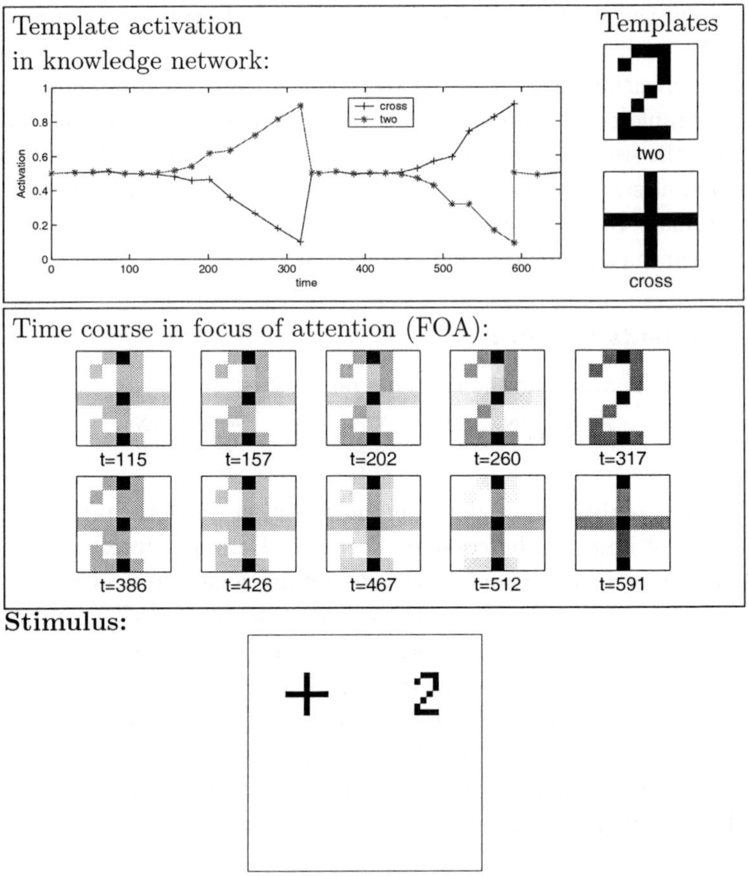

Figure 2. Basic behaviour of new SAIM.

The second term implements the second constraint:

$$E^{SN2}_{WTA}(\mathbf{y^{SN}}) = \sum_{lm}(\sum_{ij} y^{SN}_{lmij} - 1)^2 \quad (3)$$

In both terms the expression $(\sum y^{SN}_{ikjl} - 1)^2$ ensures that the activity of one location is mapped only once into the FOA.
The energy following energy function implements the neighbourhood constraint:

$$E^{SN}_{neighbor}(\mathbf{y^{SN}}) = -\sum_{i,j,l,m}\sum_{\substack{s=-L\\ s\neq 0}}^{L}\sum_{\substack{r=-L\\ r\neq 0}}^{L} g_{sr}\cdot y^{SN}_{lmij}\cdot y^{SN}_{i+r,k+s,j+r,l+s} \quad (4)$$

with g_{sr} being defined by a Gaussian function:

$$g_{sr} = \frac{1}{A} \cdot e^{-\frac{s^2+r^2}{\sigma^2}} \tag{5}$$

where A was set, so that the sum over all g_{sr} is 1. When units linked via g_{sr} are activated to $y^{SN}_{lmij} = 1$, the energy is smaller than when these units have different values, e.g. zero and one. Since g_{sr} connects units that relate to adjacent locations in both the FOA and the input image, this implements the neighbourhood constraint.
To implement inhibition of return, the location map prevents the reselection of an inhibited location through the following energy function:

$$E^{SN3}(\mathbf{y^{SN}}) = -\sum_{lmij}(1 - \sum_{lm} y^{LM}_{ij}) y^{SN}_{lmij} \tag{6}$$

The term $(1 - y^{LM}_{ij})$ suppresses already-selected locations and supports the selection of new locations.

2.5. *Knowledge network*

The energy function of the knowledge network is defined as

$$E^{KN}(\mathbf{y^{KN}}, \mathbf{y^{CN}}) = a^{KN}(\sum_{k} y^{KN}_k - 1)^2 - b^{KN} \sum_{lmn}(y^{CN}_{lmn} - w^k_{lmn})^2 y^{KN}_k \tag{7}$$

The index k refers to template units whose templates are stored in their weights (w^k_{lmn}). The term $(\sum_k y^{KN} - 1)^2$ restricts the knowledge network to activate only one template unit. The term $\sum_{lmn}(y^{CN}_{lmn} - w^k_{lmn})^2 \cdot y^{KN}_k$ ensures that the best-matching template unit is activated. a^{KN} and b^{KN} weight these constraints against each other.

2.6. *Rechecking*

In order to implement rechecking, a "location map" is computed based on activity in the selection network:

$$y^{LM}_{ij} = y^{LM}_{ij}(old) + a^{IR} \sum_{l=1}^{M} \sum_{m=1}^{M} y^{SN}_{lmij} \tag{8}$$

When a template unit in the knowledge network passes a threshold θ, the location map is used to reduce the activity in the visual field. a^{IR} controls the amount of inhibition. All units in the selection network and the knowledge network are set to the initial state they had at the beginning of the simulation.

3. Results and discussion

3.1. *Basic behaviour*

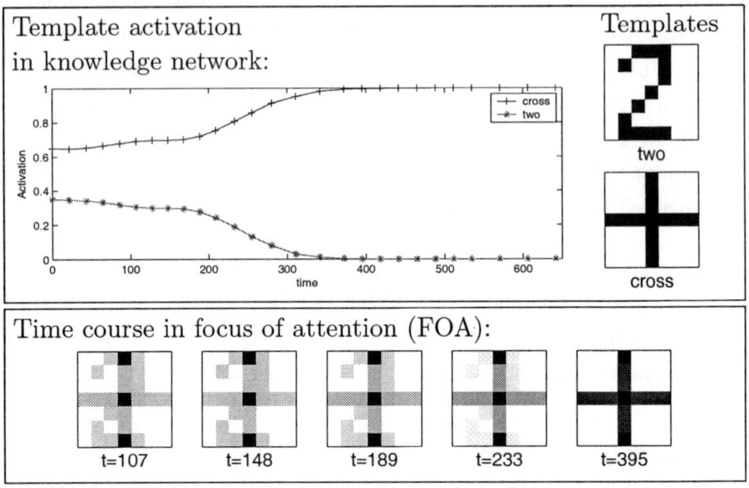

Figure 3. The simulation shows how higher initial values for the crossover comes the bottom-up bias for the two (see Fig. 2). The input image was the same as in Fig. 2

Fig. 2 demonstrates the basic behaviour of the new version of SAIM, when presented with two objects (a two and a cross). It shows that both objects are selected in a serial manner, in this case the two followed by the cross. Similarly to SAIM version 1 [1]there was a bottom-up preference towards a certain objects, here the two. This bottom-up preference results from a combination of the dynamics of the selection network and the feature extraction. We do not claim psychological plausibility for the two being preferred in selection over the cross, but it does illustrate asymmetries in bottom-up bias in the model. Fig. 3 shows that the bottom-up bias can be altered by giving the cross-template a higher initial value. This higher activation filters through the selection network via the top-down modulation (see Fig. 1). The top-down bias was used in simulations of visual search tasks where SAIM was required "to look for a target".

3.2. *Visual Search*

Fig. 4 shows that the new version of SAIM is capable of simulating typical results of visual search experiments (see [2] for examples), with linear

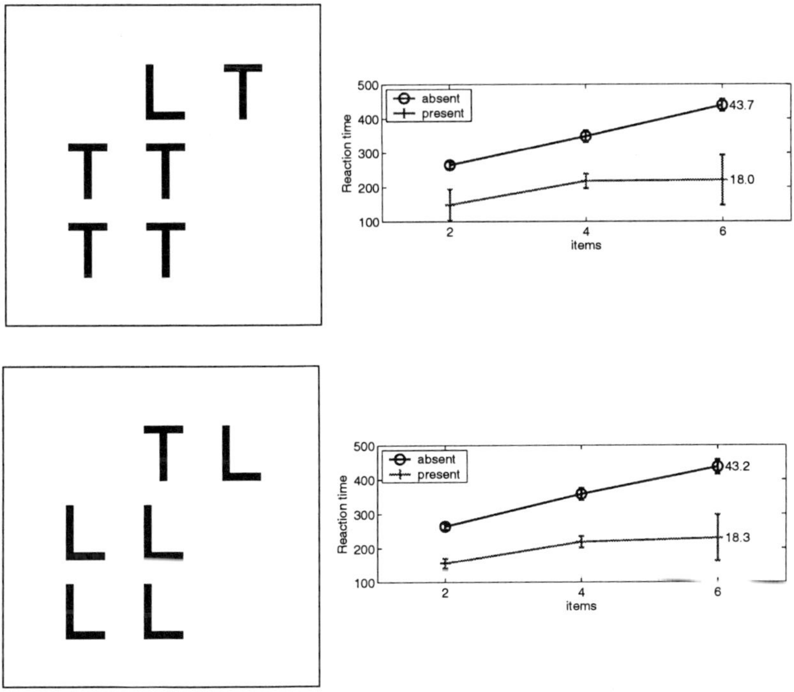

Figure 4. Two simulation results with "L" as target and "T" as distractor (top) and vice versa (bottom). The results are compatible with experimental evidence.

increases in the search functions and absent responses being lower than present responses. In these simulations each time a distractor was selected a recheck was performed with a given probability, in order to minimize target misses. Rechecking stoped entirely when either the target is found or all items were selected. The linear increase of SAIM's reaction time originates from two factors: First the time to select an item increased with the number of items, reflecting increased competition for selection. Second the number of rechecks increased with the number of items, since the probability of missing a target increases with the number of items. Fig. 5 shows the successful simulation of a search asymmetry [5], where a tilted line amongst horizontal lines is quicker than a vertical line amongst tilted lines. There is a bottom-up bias favouring the tilted line, which can only be overcome by top-down bias from the knowledge network. In SAIM the asymmetry stems from the fact that the time to select an item increases when top-down knowledge has to override a bottom-up bias. The increase results from the fact that the top-down bias from the knowledge network

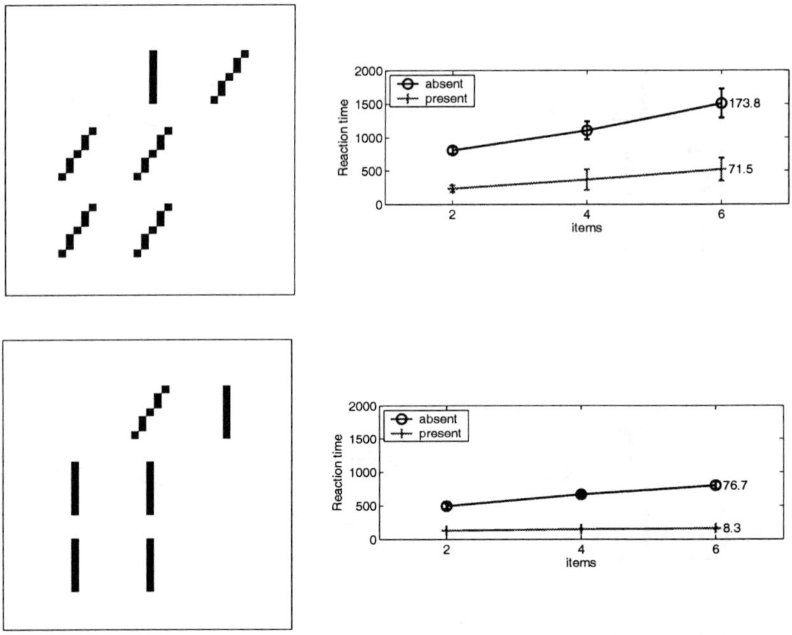

Figure 5. Simulation of a search asymmetry. Search for a tilted line amongst vertical lines is "parallel' (8.3ms/item), whereas search for a vertical line target amongst tilted lines produces a "serial" search (71.5 ms/item).

has to propagate through the contents network to the selection network. This leads to delayed resolution of activation in the selection network, after the network first follows its bottom-up preference to the distractor.

3.3. *Prediction and Experiment*

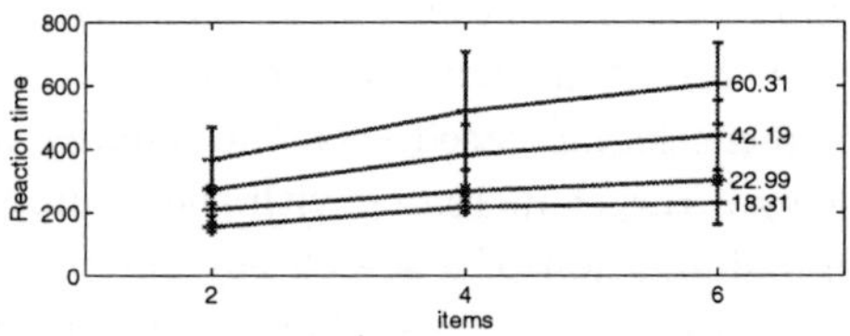

Figure 6. Effects of varying the initial activation of templates units for "T" as target and "L" as distractor. The results show that the search slope (numbers on the right to the graphs) decreased with increasing initial values.

In a series of simulations we tested the influence of the top-down bias introduced by varying the initial activation values of template units in the knowledge network. Fig. 6 shows that the search slope decreased with increasing initial activation values. This pattern results from the fact that an increasing top-down bias leads to a decrease of the selection time and an increase of the hit-rate of finding the target, reducing the amount of rechecking. We assumed that the influence of the initial value can be construed as the influence of a cue preceding search displays; the cue "sets" initial values. Search is quicker when the cue "sets" the system for the target than when the cue "sets" an alternative item. We have confirmed this prediction in an experiment on human subjects [6].

4. Conclusion

We have demonstrated that SAIM can be successfully extended to include a feature extraction process. This new version can simulate typical results from visual search experiments. It also made a prediction about effects of top-down priming of search, that we confirmed empirically. The model may provide a powerful framework for understanding human search. In future work we will aim at replacing the present feature extraction by a more biological approach (e.g. using a Gabor filter) and at capturing recent evidence suggesting that grouping interacts with attentional processes [7].

References

1. D. Heinke and G. W. Humphreys. Attention, spatial representation and visual neglect: Simulating emergent attention and spatial memory in the Selective Attention for Identification Model (SAIM). *Psychological Review*, 110(1):29–87, 2003.
2. J. M. Wolfe. Visual Search. In H. Pashler, editor, *Attention*, pages 13–74. Psychology Press, 1998.
3. J. J. Hopfield and D.W. Tank. "Neural" Computation of Decisions In Optimazation Problems. *Biological Cybernetics*, 52:141–152, 1985.
4. E. Mjolsness and C. Garrett. Algebraic Transformations of Objective Functions. *Neural Networks*, 3:651–669, 1990.
5. A. Treisman. Features and Objects: The Fourteenth Bartlett Memorial Lecture. *The Quartely Journal of Experiment Psychology*, 40A(2):201–237, 1988.
6. D. Heinke, G. W. Humphreys, and C. L. Tweed. Testing the predicition of SAIM (Selective Attention for Identification Model). *Visual Cognition*, submitted.
7. J. Driver, G. Davis, C. Russell, M. Turatto, and E. Freeman. Segmentation, attention and phenomenal visual objects. *Cognition*, 80:61–95, 2001.

TOWARDS A NEURAL NETWORK MODEL OF THE ATTENTIONAL BLINK

H. BOWMAN AND B. P. WYBLE
Computing Laboratory, University of Kent, Canterbury, Kent, CT2 7NF, UK
E-mail: {H.Bowman,B.Wyble}@kent.ac.uk

P.J. BARNARD
Cognition and Brains Sciences Unit, 15 Chaucer Rd, Cambridge, CB2 2EF, UK
E-mail: phil.barnard@mrc-cbu.cam.ac.uk

One of the most prominent experimental paradigms for investigating the deployment of attention over time is the Attentional Blink (AB). Although there is now a great deal known about it, computational modeling of the AB remains only lightly explored. This paper responds to this limitation by proposing a prototype neural network model of the blink. A central aspect of which is a realization of the concept of consolidation into working memory, which is at the heart of the majority of current explanations of the blink.

1. Introduction

The majority of visual attention research has focused on the *spatial* dimension. Despite this traditional focus, there has been an increasing amount of research directed at the *temporal* profile of attention. This research has considered how long attention is "occupied" by performing a particular task. One prominent paradigm used to investigate this issue is the *Attentional Blink* (AB) [8]. There are now many variants of the AB paradigm, but one that can claim to be canonical, and is also the one we will focus on, locates two letter targets (which we denote T1 and T2) within an RSVP (Rapid Serial Visual Presentation) stream of digit distractors [2][a]. Items in the stream are presented at a rate of approximately 10 per second and the task is to identify the two targets in a report phase that follows the stream. The characteristic empirical finding is that report of the second target is poor if it appears within a certain time interval of the first, as typically demonstrated by a serial-position curve, such as that shown in figure 2a (basic blink condition).

Although there is now a great deal known about the AB, both in respect of empirical findings, e.g. [8,2] and proposed explanations, e.g. [10], computational

[a] One reason for focusing on this experimental formulation is that it does not include a task switch, which has been argued to confound the blink paradigm [2].

modeling of the phenomenon remains only lightly explored. This paper responds to this limitation by proposing a prototype neural network model of the attentional blink. Our model is most naturally seen as a neural realization of Chun and Potter's two-stage explanation of the blink [2]. A central aspect of our model will be how it realizes the concept of consolidation into working memory, which is at the heart of Chun and Potter's and indeed the majority of current explanations of the blink. This model makes the further claim that the reason for closing the attentional gate is to allow accurate T1 binding, both in respect of binding together constituent features and binding to the correct temporal context.

The paper will begin by presenting background details on the blink in section 2. Then section 3 describes the model, section 4 presents the results of running the model and, finally, section 5 gives some concluding remarks.

2. Background on the Blink and Theoretical Justification for Model

It would clearly be impossible to do justice to the spectrum of literature concerning the AB within the context of this paper. Thus, we will simply summarize the findings that most directly impinge upon our model. Firstly, the following are key characteristics of the AB serial position curve, see figure 2a (basic blink condition),

1. the blink is a 100 to 500ms (approx) post T1 interval in which performance on T2 (conditional on correct T1 report) is significantly impaired;
2. the blink generally has a sharper onset than offset;
3. if T2 immediately follows T1 it is reported at baseline or near baseline levels (unless there is a substantial switch between T1 and T2 tasks); this is the *lag 1 sparing* phenomenon.

Modern explanations of the AB have been heavily influenced by studies that suggest that the blink has a late locus in the processing stream. Initial evidence for which came from priming studies, where it was found that with word based RSVP streams, missed T2 items primed a third target [11]. Furthermore, and perhaps even more compelling, evidence for the late locus hypothesis came from electrophysiological work, which showed that missed T2 items elicited electrical potentials associated with early perceptual activity (N1 and P2 waveforms) and with meaning (N400). However, working memory update waveforms (P3) were not present [13]. All of which suggests that the impairment to T2 processing occurs at the stage of consolidation into working memory. That is, T2s that are not reported are, broadly speaking, processed as extensively at perceptual and semantic stages as T2s that are reported, however, consolidation of their perceptual and semantic traces into working memory is prevented.

These observations beg the question of what is meant by consolidation into working memory. The theoretical perspective that we will pursue is that a central element of consolidation is *binding*. That is, in order for an item to be successfully encoded into working memory its distributed neural representation needs to be bound into a coherent whole. In fact, there are two facets of binding that we will emphasize: (1) binding together the component features of an item and (2) binding items into the correct temporal context. Theories of binding have classically focused on the first of these. However, the second is also critical since in its absence it would be impossible to recall the order in which items were encoded into working memory, e.g. for subjects to know that T1 appeared before T2 in an AB stream. While correctly ordered recall is not a requisite for accuracy in most AB paradigms and experiments, Chun and Potter [2] recorded temporal order from subjects and found order to be conserved in the vast majority of reports at lags 2 and greater. Furthermore, we would argue that even in the absence of an explicit instruction to do so, there exists a default tendency to encode and recall the order of pairs of target items that are temporally dispersed.

This binding perspective offers one reason why the gate needs to be closed at the expense of T2 accuracy. Specifically, our working hypothesis is that T2 consolidation is suppressed in order to prevent interference with T1 binding. Thus, the blink is a mechanism to ensure coherent binding of T1s. According to this explanation, lag 1 sparing is a breakdown of the system; arising because the mechanism that ensures coherent T1 binding (by closing the gate on T2) is slow relative to the Stimulus Onset Asynchrony (SOA) used in the AB paradigm. A consequence of which is that binding errors should be observable at lag 1 because coherent binding of T1 is impaired by the T2, which enters into the binding process before the gate is closed. There is some support for this theory, since in many studies T1 performance is particularly poor at lag 1, as revealed, for example, by analysis of target report percentages in table 1 of [2]. However, it could be that this lag 1 binding breakdown shows up most significantly as T1 – T2 swaps, in which featural binding of both targets is (broadly) successful, but an erroneous temporal context binding has arisen, which makes it difficult for retrieval mechanisms to correctly identify the temporal order of T1 and T2 at lag-1. Chun and Potter [2] (see figure 8 on page 119) demonstrate exactly this sort of error as do preliminary experiments conducted within in our lab. One reason for constructing the model that we present here is to investigate whether this coherent binding interpretation is consistent with the available AB data.

A final empirical finding that will greatly influence us is the observation that placing a blank after either of the targets (i.e. at the T1+1 or T2+1 positions) greatly attenuates and even in some cases eradicates the blink [8,2]. This suggests that targets are backward masked by the immediately following items [9]. That is, since they appear in the same spatial location, the iconic traces of stream items compete at a preattentive stage, with the trace elicited by a target being curtailed by the arrival of the immediately following item.

Our neural network model has also been strongly influenced by the two-stage explanation of the blink proposed by Chun and Potter [2]. In their first stage all stimuli are processed to a preliminary level at which features and perhaps even meaning are extracted. However, this level is subject to rapid forgetting and is not sufficient for report. It is only through stage 2 that stimuli are consolidated to a level required for a response. In contrast to stage 1, this second stage is capacity limited and thus, creates a bottleneck at which T2's decay while T1 is being processed

3. The Model

Our neural network model is depicted schematically in figure 1. The main layers are an *input layer* at which items are presented; a *masking layer* at which preattentive visual traces of items compete; a *category layer* at which task relevant items are foregrounded; and finally, a *working memory* mechanism, through which items are encoded and retrieved. We discuss each of these in turn. Except for the two Working Memory layers, all layers of the model use representations that are localist in nature. Thus, these layers contain one neuron for each type of item that can appear in the RSVP stream. Future extensions of the model will provide for distributed representations, but for our purposes here, simple localist representations are sufficient for testing our hypotheses.

Masking Layer. RSVP items are presented in sequence at the input layer, which feeds activation forward to the masking layer, where feedback inhibition forces these activity traces to compete. It is through this mechanism that backward (and in fact, forward) masking effects are realized. For example, in the absence of any further input, a strongly active neuron in the layer will slowly decay back to zero. However, the trace of an active neuron would be rapidly curtailed if a second item arrived at the layer during this decay period.

Category Layer. The masking layer feeds activation forward to the *category* layer. A task demand unit selectively foregrounds neurons that code target items (e.g. letters) and suppresses those that code background items (e.g. digits). Items in this layer do not compete, but their initial activation reflects the activity of stimulus traces in the masking layer. In particular, items that are masked will yield *weak traces* at the category layer and those that are unmasked will yield *strong traces*.

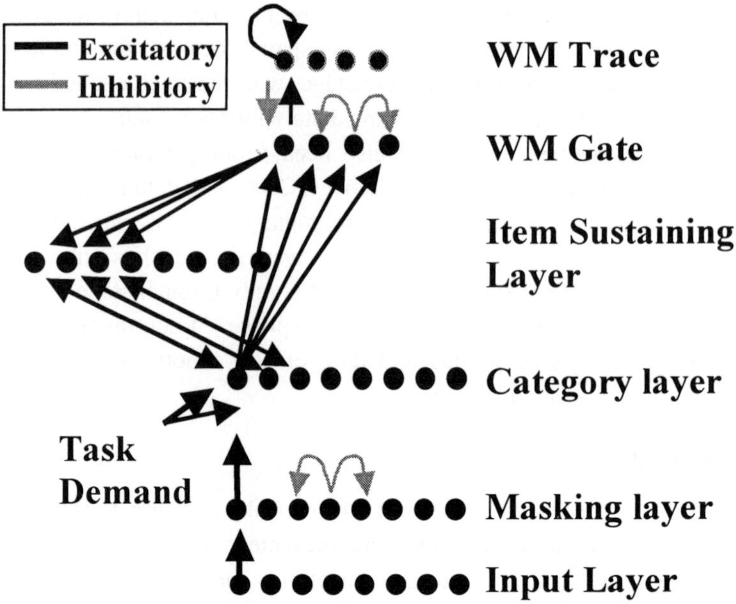

Figure 1. The Full Model. Note that the Hebbian binding links from the WM Gates to the Category layer are not included, as they play no role in the functional dynamics, serving only to indicate when successful binding has occurred.

Working Memory. As previously stated the theory that we are exploring explains the blink in terms of binding targets into Working Memory (WM). These ideas are implemented through interaction between the WM and the localist representations in the category layer. The category layer and working memory layers interact through three mechanisms. First, each category neuron directly excites all of the WM gates. Second, direct, hebbian binding links are established between specific WM gates and category layer neurons. Finally, at least one WM gate neuron has to be available to activate the item sustaining layer, which allows the system to encode, and later recall, weak (masked) items.

A pair of neurons, consisting of a WM gate and a WM trace together serve as a token [7]. Tokens are used to indicate that a target was identified, what that item was, and in what order it was perceived relative to other targets in the stream. Thus, a WM token is an encoding that combines what and when, effectively creating an "instance" of a "type". The WM gate neurons control access to their respective WM trace neurons, allowing or preventing category layer items from activating that trace. Strong winner-take-all dynamics within the gate neuron layer ensure that only one trace neuron is accessible at a time. Each trace neuron is individually self-excitatory and can self-sustain for the duration of a trial once activated. Working memory consolidation is implemented in terms of building *binding links* between WM gates and category layer neurons. Thus, an item is viewed as having been consolidated into working memory if a link is successfully built from a WM gate to the neuron coding that item at the category layer. Conceptually, these links should be viewed as pointers from a given WM *token* to the featural / categorical neural circuits that code the *type* of the item being consolidated, which enable later top-down retrieval of that item. Binding links are built via a rapid hebbian process between WM gates and category layer neurons[b]. These links are unidirectional and play no part in the functional dynamics of the model during the presentation of the input.

In addition to WM neurons serving as pointers in order to enable later retrieval, they also code temporal order. This is obtained by, firstly, using a winner-take-all mechanism amongst WM gates to ensure that only one gate neuron is active at any instant; this neuron denotes the current temporal context. Furthermore, each WM trace neuron sends an inhibitory projection to its corresponding gate neuron. This ensures that when a WM trace neuron has been activated, it closes its own WM gate for the remainder of the trial. The closing of one such gate initiates winner-take-all competition amongst the remaining WM gate neurons until a new gate is made available for future binding.

Thus, gates are made available in sequence in an order determined by the bias inputs applied to them. In "normal" processing, i.e. when the time gap between pairs of (to be consolidated) targets is long, each WM gate obtains a link to a single category layer item, denoting that a different temporal context has been allocated to each recognized target. However, when target items arrive at the category layer in close temporal proximity (as arises at short lags in AB streams), the handover between WM gates can be too slow to keep up. Consequently, binding errors can occur, which arise in the model when a single WM gate obtains links to multiple category layer items. In addition, T2s can be missed altogether because their category layer activation falls in the window between one WM gate (which has encoded T1) being suppressed and the next one becoming available.

Item Sustaining Layer. The final mechanism that we need to explain is the *item sustaining layer*. As previously discussed, masking plays an important role in obtaining the blink. This is reflected in our model since binding links can only be constructed between strongly active WM gates and strongly active category

[b] This use of Hebbian learning might be viewed as controversial, however, it is not essential to our model and could be replaced by an activation-based gating mechanism.

neurons (this is built into our Hebbian learning rule). Thus, in the absence of further stimulation, binding fails for category layer traces of masked targets.

Conceptually, we believe that the brain provides a mechanism to "grab hold" of such weak, but *task relevant*, stimuli and sustain them for a sufficiently long period that they can be bound into working memory. The item sustaining layer implements such a mechanism by providing a recurrent excitatory circuit to prolong the duration of traces in the category layer neuron in order that they can be encoded (similar techniques can be found elsewhere, e.g. [4]). For this excitatory circuit between a given category neuron and its dedicated sustaining neuron to be active, concurrent input from both the category layer and a WM gate must be present. Since each WM gate connects to all of the sustaining neurons, any one active gate will enable all of the sustaining neurons. As the network completes the process of encoding the T1, the system undergoes a switch from one active WM token to the next. During this handoff process, all of the WM gates are inactive and consequently all of the sustaining neurons follow suit. It is during this handoff from one token to the next that the system is no longer able to encode masked items because of the inactivity of the item sustaining layer. It is imperative that the item sustaining layer is temporarily shut down in this way or the first item of any sequence would be encoded to all of the available WM tokens. The subject would recall multiple instances of an item that was presented only once. This scenario is strongly contra indicated by data on the repetition blindness effect [7].

The length of time required by this system to encode an item is determined by the strength of that item. Strong (unmasked) items are rapidly encoded by strongly activating WM gates, which causes binding links to be built quickly. Conversely, weak items bind more slowly and require prolonged assistance from the item sustaining neurons.

This is the heart of how the model blinks, yet exhibits lag 1 sparing. A T2 presented immediately after a T1 has a chance of building a binding link with the same WM gate neuron. While this is technically an error, we propose that the system is able to disambiguate this double-encoding during retrieval in a process not explicitly modeled here. We further propose that it is because two items are bound to the same token that it is difficult to correctly recall the order of these items. This has been shown by experimental work of Chun and Potter [2] who demonstrate a selective impairment in the recall of temporal order at lag 1.

If the SOA between a masked (and therefore weak) T1 and T2 is between 200 and 400 ms, the WM door will close prior to creation of the binding link to the T2 and the category trace will fade and be lost during the switch from one WM gate to the next. In this way the door is closed on T2 while T1 is being bound into WM and the length of time it takes T1 to be bound regulates the length of the blink. Unmasked (and therefore strong) T1's are able to establish binding links with the appropriate WM neuron more rapidly. Thus, the blink is shorter and shallower when T1s are unmasked rather than masked.

A MATLAB implementation of this model is available at www.cs.kent.ac.uk/people/staff/bw5/ncpwblinkmodel/.

4. Results

The crucial performance measure of the Attentional Blink paradigm is the successful encoding of T2 for trials in which T1 was encoded (T2|T1). In this model, encoding of a target was scored as successful if the binding links from any WM gate to that target were above a designated threshold. Thus, we have not at this stage considered how the number of T1 – T2 swaps varies with lag, although this information could easily be extracted from our model. The elements of the conventional AB paradigm that will be modeled explicitly in this paper include the deficit at lags 2-5, the relative sparing of performance at lag 1, and the attenuation of the blink curve by blank(s) following T1 and T2.

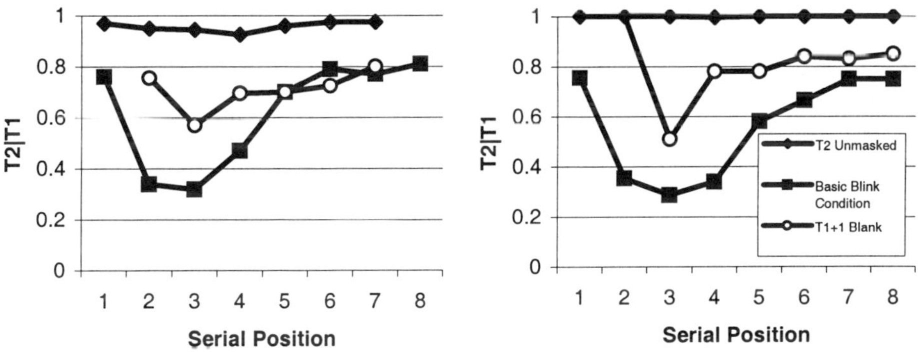

Figure 2. (a) Human data (on left) and (b) model data (on right).

The data being modeled will be extracted from [2] and [6]. Both of these papers used paradigms that are generally equivalent to the one used here. SOA was 100ms and targets were digits in a letter stream. T1 and T2 tasks were identical. This data is available from the aforementioned papers for the three conditions studied: basic blink condition [2], T1+1 blank [2] and T2 as the final element of the stream [6]. Figure 2a presents these three conditions for the experimental data while Figure 2b presents data for the same three conditions in the model. It should be clear from these results that, at least in qualitative terms, our model successfully reproduces these three experimental conditions.

The elevated performance for lag 2 in the T1 +1 Blank condition is due to a lack of forward masking for the T2 item, which increases its strength, compounding the attenuated blink caused by the stronger T1 trace. The experimental data from Chun and Potter [2] indicates the same sort of effect at lag 2, although to a lesser degree. In our model, any manipulations that increase the strength of T1 and T2 items attenuates the blink. Therefore this forward masking effect, inherent in the design of the masking layer, allows our model to make the prediction that the attentional blink will be attenuated by blanks at positions T1-1 and T2-1.

5. Conclusions

We have presented a prototype neural network model of the AB, which has enabled us to explore how key AB data can be reproduced by a WM consolidation model. According to this theory, in order to protect T1 from binding errors, a door is closed on the consolidation of T2 targets if their category layer trace falls within the window of T1 binding. This is the blink window. However, the door is not closed instantaneously. A result of which is that lag 1 T2s can be consolidated. However, this process can erroneously bind T1 and T2 to the same WM token. This error allows T1 and T2 to be retrieved at lag 1, but without their correct temporal order. In addition, the model reproduces AB masking effects. Specifically, the blink is attenuated if either the T1+1 or the T2+1 items are left blank. In the former case this arises because strong T1s are rapidly consolidated into working memory and thus, a fresh WM token is released before T2 has decayed, while in the latter case the blink is attenuated because stronger T2s can out-live the blink window.

It is beyond the scope of this paper to give a detailed comparison of our model to existing theories of working memory and prefrontal function. However, it is safe to say that our model has similarities to a number of such theories, e.g. in respect of foregrounding task relevant items, c.f. [3,1]; allocation of general purpose WM resources (our WM gate and trace neurons), c.f. Duncan's adaptive coding theory [5]; sustaining activation by setting up reverberating circuits, c.f. Dehaene et al's global workspace resource [4]; and active maintenance which occurs in our WM trace neurons, c.f. [12].

6. Acknowledgements

This work is supported by a grant from the UK EPSRC (GR/S15075/01).

References

1. Braver, T.S. and J.D. Cohen (2000) "On the Control of Control: The Role of Dopamine in Regulating Prefrontal Function and Working Memory" Control of Cognitive Processes, Attention and Performance XVIII, S. Monsell and J. Driver (eds), MIT Press, 713-737.
2. Chun, M. M. and M. C. Potter (1995). "A Two-Stage Model for Multiple Target Detection in Rapid Serial Visual Presentation." J Exp Psych: HPP **21**(1): 109-127.
3. Cohen, J.D., K. Dunbar, et al. (1990) "On the Control of Automatic Processes: A Parallel Distributed Processing Account of the Stroop Effect" Psychological Review **97**(3): 332-361.
4. Dehaene, S., M. Kerszberg et al (1998) "A Neuronal Model of a Global Workspace in Effortful Cognitive Tasks" Proc Natl Acad Sci USA, **95**: 14529-34.
5. Duncan, J. (2001) "An Adaptive Coding Model of Neural Function in Prefrontal Cortex" Nature Reviews, Neuroscience, **2**, Nov., 820-829.
6. Giesbrecht, B. and V. Di Lollo (1998) "Beyond the Attentional Blink: Visual Masking by Object Substitution" J Exp Psych: HPP **24**(5): 1454-66.
7. Kanwisher N. (1991) "Repetition blindness and illusory conjunctions: errors in binding visual types with visual tokens." J Exp Psych: HPP **17**(2):404-21.
8. Raymond, J.E., K.L. Shapiro et al (1992) "Temporary Suppression of Visual Processing in an RSVP Task: An Attentional Blink" J Exp Psych: HPP 18(3):849-860.
9. Seiffert, A.E. and V. Di Lollo (1997) "Low-Level Masking in the Attentional Blink" J Exp Psych: HPP **23**(4): 1061-1073.
10. Shapiro, K. L., K. M. Arnell, et al. (1997). "The Attentional Blink." Trends in Cognitive Sciences **1**(8): 291-296.
11. Shapiro, K. L., J. Driver, et al. (1997). "Priming from the attentional blink: A failure to extract visual tokens but not visual types." Psych Sci **8**: 95-100.
12. Usher, M. and Davelaar, E. (2003) "Short/long term memory in terms of activation versus weight based processes" this volume.
13. Vogel, E.K., S.J. Luck et al (1998). "Electrophysiological Evidence for a Postperceptual Locus of Suppression During the Attentional Blink" J Exp Psych: HPP **24**: 1656-1674.

LIMITED CAPACITY DIMENSIONAL ATTENTION AND THE CONFIGURAL-CUE MODEL OF STIMULUS REPRESENTATION

PAUL D BARTOS

Psychology Department, The Open University, Milton Keynes, MK7 8AA, United Kingdom

Gluck and Bower's[1] configural cue model is a network that represents stimuli using independent nodes for each feature and feature combination within the stimulus. One of its main limitations is the lack of any clear method for incorporating secondary learning processes such as selective attention. A new configural cue model is proposed in which node activation is dependent on the average characteristics of a dimensional sampling process. This process may be described in terms of a Markov process. Learning algorithms are used to alter the matrix of transition probabilities governing the behavior of the sampling process on each trial. This allows the model to qualitatively simulate learning effects that seem to be based on limited-capacity dimensional attention. The approach used also allows the model to be used to simulate attention learning and associative learning with feature-based stimuli. This represents a potential advance over many models used in category learning research where dominant models are either only applicable to stimuli that do not vary in terms of their dimensionality (such as ALCOVE[2]), or make use of stimulus representations that are incapable of learning nonlinear discriminations (such as EXIT[3]).

1. Introduction

One difficulty that obtains in connectionist modeling of category learning is that the models that can accurately simulate learning about dimensional stimuli (stimuli with fixed dimensionality, varying from one another in terms of the values they take on each dimension), are not well suited to simulating learning where stimuli consist of collections of unrelated components.

Two sets of experimental findings will be examined, one concerning dimensional stimuli, and one noted in studies using feature-based stimuli. No model with a single form of stimulus representation can be used to simulate both sets of findings. As an approach to addressing this problem, a new model is proposed that makes use of the configural cue form of representation.

The configural-cue model represents stimuli in terms of the power set of their components. A stimulus with n components will be represented by the activation of 2^n-1 nodes, one node for each component and compound possible given the number of dimensions. The activation of these representations is usually binary, active if the feature or feature set it represents is present, and inactive otherwise.

A substantial body of experimental findings support the idea that human performance in category learning experiments is mediated by 'selective attention' processes that vary the utilization and associability of stimulus elements or whole stimulus dimensions. This variability seems to depend on participants learning and remembering the relative relevance of elements or dimensions to task performance across learning trials[4,5,6,7,3]. One of the configural cue model's main limitations, with respect to simulation, is a lack of any way of implementing this kind of selective attention.

The new model proposed here affords the configural cue model a form of selective attention. Dimensionality is imposed on the configural cue model by grouping representations according to their membership of spaces. Measures relating to the spaces control activation and learning rates for members of that space. Where stimulus components do not share and vary in terms of salient dimensions, components are each regarded as dimensions in themselves.

The main purpose of the model is to provide a method of implementing selective attention for both dimensionally constructed stimuli and for feature-based stimuli. The model is later tested on one experiment in which the effects of dimensional relevance on task difficulty are examined. It is then examined in relation to a generalization effect noted with feature-based stimuli.

1.1. *Dimensional attention*

In their classic study of category learning rates, Shepard *et al.*[4] required participants to categorize eight, patterns or objects, varying in terms of positions on the same three dimensions, into two groups of four according to one of six experimenter-defined category types. The abstract structures of these types are shown in figure 1.

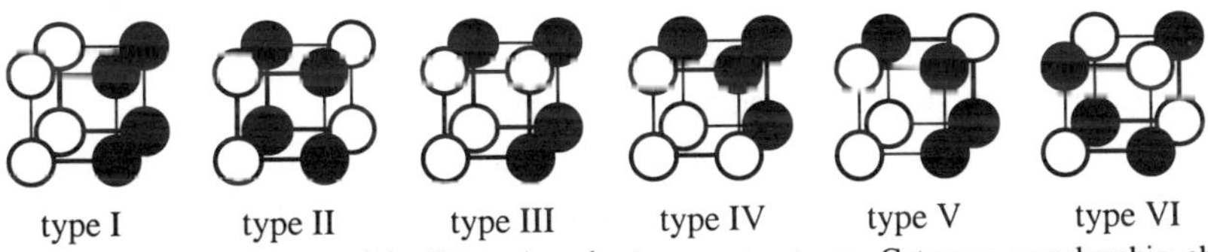

Figure 1. Abstract structure of the Shepard *et al.* category structures. Category membership shown by filled or open circles. In the text left-to-right dimension is indexed 'q', bottom-to-top is 'r', and front-to-back is 's'.

Type I requires knowledge of just one (q) dimension to successfully predict the category label. Type II is an XOR task in which knowledge of two dimensions, q and s, is required. Types III to V are rule-plus-exception structures where all three dimensions are relevant but to different extents. For type VI, knowledge of all three dimensions is required for all stimuli.

The results of these experiments revealed systematic differences in the learning rates for different category structures. The type I was the easiest, followed by the type II. Types III to V were next, being of approximately the same difficulty, with the type VI hardest of all. The fact that ease of learning relates to the number of dimensions unnecessary for the task has lead to the proposal that some form of limited capacity dimensional selective attention is involved in this kind of learning[4,5].

Nosofsky *et al.*[5] tested a number of connectionist models that had shown some success at predicting other features of category learning on these tasks. The most successful model in these tests was ALCOVE[2,8], which combines connectionist learning rules with the exemplar representations of the context model[9] and general context model[10,11]. ALCOVE uses a form of back-propagation of error, to control the capacity-limited sensitivity to differences in dimensional input. The configural cue model, like basic exemplar models[5], tended to learn the type III task more quickly than the type II. The model displays very similar performance on types II, III, IV, and V.

Note that the exemplar model of representations was developed to account for generalization in continuous *fixed* dimensionality spaces. It does not make coherent statements about learning and generalization when the number of dimensions in the stimuli varies as it does for feature-based stimuli.

1.2. *The inverse base-rate effect*

The inverse base-rate effect is noted in experiments involving feature-based stimuli paired with outcomes with different frequencies[12,7,3]. The basic effect can be described in terms of two stimuli and two categories. On 0.75 of learning trials the compound to category pairing AB→1 is presented. On the remaining 0.25 of the trials the compound BC→2 is presented. Following training on this set, the participant is presented with the compound AC. On average, participants tend to assign the novel compound to the rare category 2.

Kruschke's theory[7] regarding the effect is that because participants are generally exposed to the common pairing first, they will distribute associative strength equally across its two components. When exposed to the rare stimulus for the first time, participants rapidly shift attention away from error-increasing components and thus allocate the bulk of BC's associative strength to the C component. When A and C are presented in the test compound, C's weights should generally be larger than A's, so the rare category 2 will be in receipt of more response strength.

The models that Kruschke proposes to simulate the effect, ADIT[7] and EXIT[3], have rapid attention shift algorithms to implement the theory described above. Feedback is followed by a fixed number of iterations of an attention learning back propagation algorithm. Associative weights are then updated at the end of the trial. It is important to note that these models use elemental or component cue representations with one node for each stimulus component. This makes these models incapable of simulating human learning of simple nonlinear discrimination problems, for example learning to associate a compound AB with one category and learning to associate individually presented components A and B with different categories. Note that structures II to VI of the Shepard *et al.* experiment are also nonlinear discrimination tasks and as such, these models are unable to simulate the fact that people can master all six of these structures.

2. Model description

2.1. *Structures and algorithms*[a]

Stimuli are represented using the configural cue form of representation (see section 1). Associative weight is delivered to the decision function at a rate given by the product of two measures. The first measure is the presence (1) or absence (0) of the dimensional events that the node represents. The second activation measure is the 'sampling rate' for the space.

Activation is dependent on a dimensional sampling process. This process is represented by the average behavior of a markov process, with each dimension a node with a base probability generally equal to one over the number of present dimensions. The probability of transition between dimensions is governed by a matrix of transition probabilities. The activation of a single-dimensional representation is given by the average sampling rate for its dimension, given the base probabilities and the transition matrix. The activation of a multidimensional representation is given by the average rate at which the process samples each of its component dimensions *consecutively*. So, for example, the sampling rate for a 2d space, qr, p(qr), is given by $p(q)p(r|q) + p(r)p(q|r)$, where p(q) and p(r) are the average sampling rates for dimensions q and r.

Calculating the sum of probabilities for activating sequences for a particular space requires two sets of values. The first is the average sampling probabilities for each dimension in that space. The second is the conditional or transition probabilities between each space dimension r at markov step μ and each space dimension q at step $\mu+1$.

The average sampling rates for each dimension are given by multiplying the matrix of transition probabilities at iteration t, $\mathbf{P}_t$, raised to some power, M, (8

---

[a] For a full description of the model, contact paul.bartos@virgin.net

used) by the vector of base probabilities $\mathbf{p}'$ to yield a vector of dimensional sampling probabilities $\mathbf{p}_t$;

$$\mathbf{p}'\mathbf{P}_t^M \underset{M \to \infty}{\rightleftarrows} \mathbf{p}_t . \qquad \text{(Eq.1)}$$

The base probability of a dimension q, p′(q), is generally given by its presence (one or zero) over the sum of presences for all dimensions. The matrix of conditional probabilities that applies at iteration t, $\mathbf{P}_t$, for each origin dimension r at markov step μ and each destination dimension q at step μ+1 is calculated by the following:

$$p\left(q_{\mu+1} \mid r_\mu\right)_t = \frac{p'(q)_t \, a_{rq(t)}}{\sum_Q p'(Q)_t \, a_{rQ(t)}} . \qquad \text{(Eq.2)}$$

The contribution of the learnt transition weights is indexed $a_{qr(t)}$. This is given by the product of the presence of the two dimensions multiplied by a logistic transform of the sum of long-term and within-trial transition weights. During the learning phase of the trial the long-term weight is multiplied by the choice probability for the correct output. Transition weights begin at zero and may become positive or negative, dependent on whatever learning takes place.

A learning trial is simulated by calculating the base probabilities and transition matrix applicable to the current stimulus. From this, the outputs and the choice probabilities are calculated. There follows a number of iterations of the learning process. Panel A of figure 2 shows the steps involved in the model's learning process.

Attention learning is implemented in this model by comparing multiple 'perspectives' on the relationship between the transition matrix and the output. This algorithm uses the sum of associative strength delivered to the current category at iteration t, $O_{j(t)}$, as a baseline upon which transition weight changes are based. If this value is greater than or equal to unity, no attention learning occurs; otherwise the algorithm generates two output-based measures with respect to each present dimension. The first is generated by increasing the base probability for the dimension and recalculating the sampling rates, outputs, and error measures. The second is generated by repeating the process but with the base probability for the dimension reduced. Apart from the new base probabilities, the values involved in the generation of the transition matrix are held constant for both measures for all dimensions.

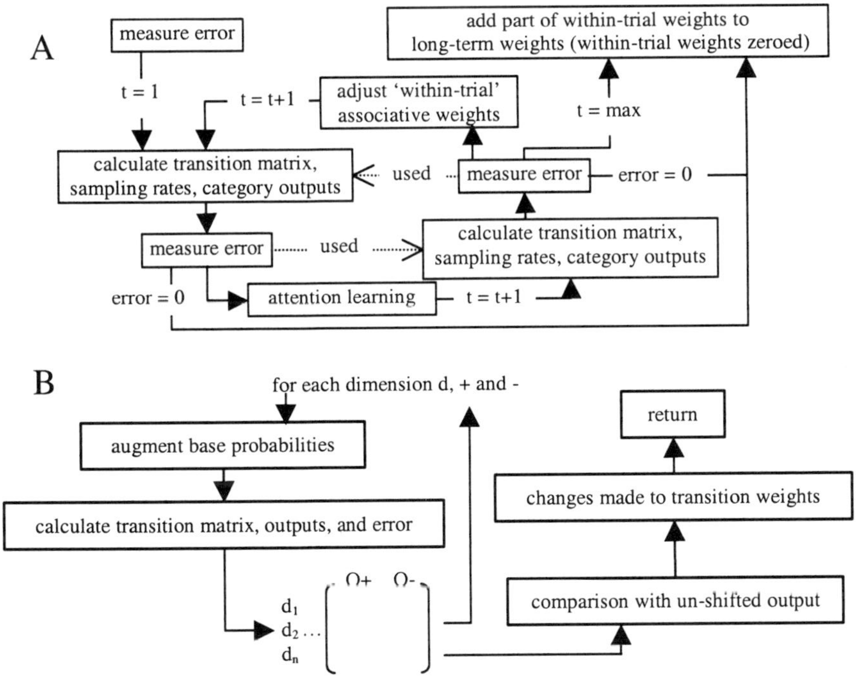

Figure 2. A) Cycle and sequence of events in the model's learning algorithm. The learning process begins at the end of t= 0 when feedback is presented about the current stimulus. B) Operation of the multi-perspective attention-learning algorithm.

The change to a within trial recurrent weight, for a dimension, q, is calculated by subtracting the baseline $O_{j(t)}$ from the value of $O(q+)_{j(t)}$ (the measure obtained when the base probability for q was increased). The change for a transition weight between two different dimensions, e.g. from q to r is given by $O_{j(t)}$ - $O(r-)_{j(t)}$. Added to this is $O(r+)_{j(t)}$-$O_{j(t)}$ minus $O(q+)_{j(t)}$-$O_{j(t)}$. If ignoring a dimension results in lower output to the correct category, transition weights towards it will increase. This relationship is augmented using the difference between the recurrent weight signal for the destination and that for the origin. Change signals for both types of within-trial transition weights are multiplied by 1- $O_{j(t)}$ and a positive constant.

Associative weights are altered by adding a teacher signal, multiplied by the presence of the representation, the sampling rate for its space, and a positive constant. The teacher signal is given by the feedback to a category minus the summed output to that node unless the output exceeds the feedback and has the same sign. In this case the teacher signal is zero[2].

Learning stops after a fixed number of iterations, or if the total error is reduced to zero. At the end of learning, the long-term transition and associative weights are adjusted by adding to them their within-trial weights multiplied by learning rate parameters.

3. Simulations

3.1. *Dimensional selective attention*

The experiment used a repeated presentation-guess-feedback cycle. Participants were trained across 25 blocks or until they reached a criterion of no errors for 32 trials. Each block consisted of 16 trials, each object presented twice with order randomized[4,5]. The model simulated this by generating 30 trial sequences according to the experimental design. The model simulated this by generating 30 trial sequences according to the experimental design.

The model uses six adjustable parameters. The within-trial and long-term transition weight learning rates were both set at 0.75; the within-trial and long-term associative weight learning rates were set at 0.4 and 0.5 respectively. The maximum number of iterations for the learning process was 40 and the constant used to increase and decrease base probabilities in attention learning was 10.

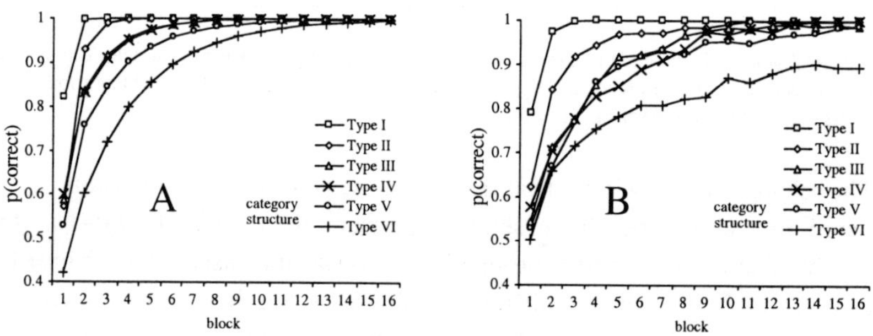

Figure 3: Model simulation (panel A) and human data (from Nosofsky *et al.*[5]) (panel B) for mean probability of correct response per block of training.

Mean probability of correct responding, per block on each of the six category structures (see figure 1) is shown in figure 3. Panel B of this figure shows the average learning curves for Nosofsky *et al*'s[5] human participants. As can be seen from figure 2, the model presented here provides a qualitative fit to the human data with the observed order of difficulty simulated.

Table 1 shows the mean space sampling rates at t = 0 across the seven spaces for the last block of training on each of six tasks. Table 1 shows the mean space sampling rates at t = 0 across the seven spaces for the last block of training on each of six tasks. For the type I task, a transition matrix has emerged that strongly enhances the activation of the single essential q dimension. This is implemented by a high recurrent q-to-q weight and high transition weights towards q from other dimensions. All other weights will be strongly negative. The type II transition matrix strongly enhances the sampling rate for the essential qs space. Here, q-to-q, s-to-s, q-to-r, and s-to-r weights will be highly negative with weights between q and s being highly positive. For the remaining tasks, the transition matrix is promoting a highly distributed sampling process favoring the activation of the qrs space.

Table 1. Mean sampling rate for each space (sum of probabilities of space-activating sampling sequences) for last block of training on the six category structures. Essential dimensions and spaces for types I, II, and VI structures shown in bold.

	Category structure					
space	Type I	Type II	Type III	Type IV	Type V	Type VI
q	**0.92**	**0.49**	0.33	0.33	0.33	0.33
r	0.04	0.03	0.33	0.33	0.33	0.33
s	0.04	**0.49**	0.33	0.33	0.33	0.33
qr	0.05	0.03	0.33	0.33	0.33	0.33
qs	0.06	**0.91**	0.33	0.33	0.33	0.33
rs	0.02	0.02	0.33	0.33	0.33	0.33
qrs	0.03	0.07	0.49	0.5	0.5	**0.5**

3.2. *The inverse base-rate effect*

The test for the model was a considerable simplification of Kruschke's[7] experiment 1. The model was tested with two stimuli and two categories, across six training blocks. One stimulus, the compound AB is presented paired with category 1 on three of the four trials in each block. The other stimulus, the compound BC, is presented once per block, paired with category 2. Kruschke's original experiment involved participants learning two of these base-rate structures with four stimuli and four categories spread across 15 blocks of eight trials each. The data presented in table 2 is from Kruschke's[7] experiment 1 but the probabilities of selecting a category from the wrong substructure are ignored and data are collapsed across both substructures

The within-trial and long-term transition weight learning rates were set at 0.75 and 1 respectively; the within trial and long-term associative weight learning rates were set at 0.3 and 0.4 respectively. The maximum number of iterations for the learning process was 40 and the constant used to increase and decrease base probabilities in attention learning was set to 10. Thirty randomized input sequences were generated.

Table 2. Comparison of human and model average response probabilities for five transfer stimuli. Human data is from Kruschke[7], experiment 1 collapsed across substructures (see text).

	human		model	
stimulus	p(1)	p(2)	p(1)	p(2)
A	0.968	0.032	0.864	0.136
B	0.811	0.189	0.576	0.424
C	0.042	0.958	0.033	0.967
AC	0.366	0.634	0.419	0.581
ABC	0.591	0.409	0.470	0.530

The inverse base-rate effect with respect to the AC compound is demonstrated by the model. Because the rare compound generally occurs after the common compound has been presented, the attention-learning algorithm tends to result in strong positive weights from C to C and from B to C, as enhancing the base probability for C will reduce error. C thus tends to get larger associative weight changes than either A or B and, in addition, tends to be recurrently sampled more frequently than A. In addition, a significant proportion of the associative strength for the AB compound will be dependent on the activation of the AB configural representation, which does not occur when the AC compound is presented. The model predicts generally equal response probabilities for the ABC transfer compound, at variance with the human data that suggests that the common category will be favored. Kruschke's[7] theory about the inverse base-rate effect, discussed in section 1.2, does not have anything to say about this particular transfer stimulus. Further research seems to be required to generate theories as to why a base-rate consistency is exhibited on presentation of this stimulus.

4. General discussion

The model presented here seems to allow configural cue representations to be used in simulations of attention learning effects with both dimensional and feature-based stimuli. The attention-learning algorithm seems to capture some key effects of learning and performance with both types of stimuli. At the moment, there are no models capable of offering any account of learning where stimuli might vary in terms of dimensions, but also in terms of the presence and absence of those dimensions.

One shortcoming of the form of representation used here is that, like the basic configural cue model, it is unable to simulate generalization across continuous dimension spaces. This is a clear advantage for the exemplar model. Given that the exemplar model is incapable of offering accounts of learning about feature based-stimuli, however, further research may be appropriate to investigate whether and how a configural cue form of organization may be adapted to simulate continuous dimension generalization.

References

1. M. A. Gluck and G. H. Bower. Evaluating an adaptive network model of human learning. *Journal of Memory and Language*, **27**, 166-195 (1988).
2. J. K. Kruschke. ALCOVE: An exemplar-based connectionist model of category learning. *Psychological Review,* **99**, 22-44 (1992).
3. J. K. Kruschke. Toward a unified model of attention in associative learning. *Journal of Mathematical Psychology,* **45**, 812-863 (2001).
4. R. N. Shepard, C. I. Hovland and H. M. Jenkins. Learning and memorization of classifications. *Psychological Monographs*, **75** (13, Whole No. 517) (1961).
5. R. M. Nosofsky, M. A. Gluck, T. J. Palmeri, S. C. McKinley and P. Glauthier. Comparing models of rule-based classification learning: A replication and extension of Shepard, Hovland, and Jenkins (1961). *Memory and Cognition*, **22**, 352-369. (1994).
6. N. J. Mackintosh. A theory of attention: Variations in the associability of stimuli with reinforcement. *Psychological Review,* **82**, 276-298 (1975).
7. J. K. Kruschke. Base rates in category learning. *Journal of Experimental Psychology: Learning, Memory and Cognition*, **22**, 3-26 (1996).
8. J. K. Kruschke. Human category learning: Implications for backpropagation models. *Connection Science,* **5**, 3-36 (1993).
9. D. L. Medin and M. M. Schaffer. Context theory of classification learning. *Psychological Review,* **85**, 207-238 (1978).
10. R. M. Nosofsky. Choice, similarity, and the context theory of classification. *Journal of Experimental Psychology: Learning, Memory, and Cognition,* **10**, 104-114 (1984).
11. R. M. Nosofsky. Attention, similarity and the identification-categorization relationship. *Journal of Experimental Psychology:General,* **115**, 39-57 (1986).
12. D. L. Medin and S. M. Edelson. Problem structure and the use of base-rate information from experience. *Journal of Experimental Psychology:General,* **117**, 68-85 (1988).

High Level Cognition and Implementation Issues

A TEMPORAL ATTRACTOR FRAMEWORK FOR THE DEVELOPMENT OF ANALOGICAL COMPLETION

R. LEECH, D. MARESCHAL AND R. COOPER
School of Psychology, Birkbeck, University of London
Malet Street, London WC1E 7HX, UK
E-mail: r.leech@psychology.bbk.ac.uk

The current model is an adaptation of [1], extending it to draw more complex and abstract analogies. Units are connected by two types of modifiable connections: fast connections which transmit the current activation of the units and slow connections which implement a delay transmitting an earlier activation state of the network. The fast connections drive the network into attractor states corresponding to objects. The slow connections implement transformations between states by pushing the network out of its stable state and into another attractor basin. The fast and slow connections work together to move the network from one attractor state to another in an ordered way. Since the network can learn transformations between more than two objects we suggest how the network could draw analogies involving more than two objects.

1. Introduction

Analogy is widely held to be a central cognitive process, both in adults and in children [2]. Children use analogy, often spontaneously, to solve problems and to learn about biological, physical and psychological domains [3]. Even children as young as 10-months old have been observed to use analogy in problem solving [4]. However, despite its cognitive importance, existing mechanistic accounts of the development of analogy have tended to treat it as an offshoot of adult analogical reasoning (e.g. [5]). In contrast, [1] sets out a purely developmental account of the emerging ability to draw simple analogies. Under this account analogy occurs, not as the result of any special mechanism, but as the result of gradual accumulation of factual knowledge in a simple memory system. The current paper is an attempt to build on this earlier work by adapting [1] to incorporate, more fully, temporal sequences of events. We argue that this approach has the potential to account for a wider range of analogies.

The rest of this paper unfolds as follows. First we briefly review the connectionist account of [1], considering both the network and the analogy it simulates. Then we highlight several reasons for adapting the model, before detailing the new framework. We see how the extended model solves the same analogies as [1] before suggesting future directions.

1.1. *The earlier model*

The earlier model was designed to complete a specific subset of analogies involving casual transformations. [6, 7]. This task involves A is to B as C is to ...? analogies, familiar from IQ tests, which have been designed to be accessible to young children. The analogy involves pictures of familiar objects which may have been causally transformed by a very familiar relation, e.g. *cutting* or *melting*. A typical analogy might be, presenting the child with a picture of bread then a picture of a cut piece of bread before presenting the child with a picture of an apple. The child then has to choose from amongst four pictures (the correct pictures and three distractors) the appropriate response to complete the analogy (in this case the picture of cut apple).

Both in the present work and in [1], analogy completion is driven by priming. For an A:B::C...? analogy, the A and B terms prime a relation which then biases the C term to produce the analogically appropriate D term. The development of the analogical abilities is achieved through the gradual acquisition of knowledge about relations. As with [1], the current account differs from most other models of analogy (e.g. [9]) by avoiding an explicit structure-mapping system or any built-in syntactic structures, such as predicate calculus. Instead, [1] and the current work suggest that analogy might be explained by far simpler memory-based mechanisms.

The model learns causal relational information from which analogies can be formed. A causal relation is assumed to be embodied in the transformation of state A to state B (e.g. the transformation of "apple" to "cut apple"). The similarity between two different examples of a relation (e.g. the similarity between apple:cut-apple and bread:cut-bread) lies in how the perceptual features of the objects involved are transformed.

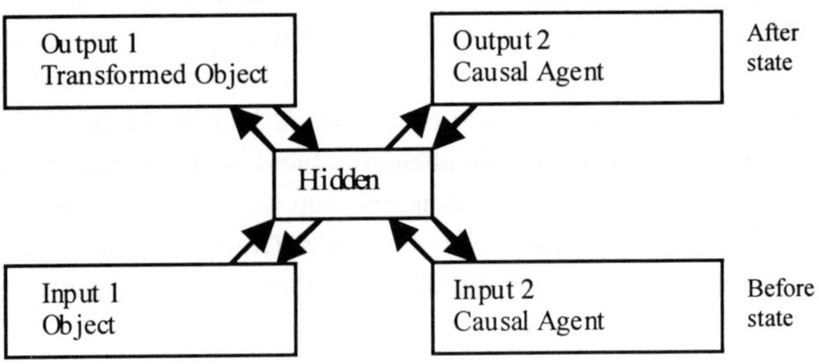

Figure 1. Schema of the earlier model architecture. The input layer codes objects before a transformation, and the output layer codes objects after a transformation has occurred.

All network weights are bi-directional and symmetrical. The input layer represents the "before" state of an object and the output layer represents the "after" state of the same object. Both the input and output layers are split into two, one representing an object which could be in a "before" state or a different "after" state, and a causal agent. The causal agent could be thought of as the situation or object that leads to the transformation, i.e. for *cutting* the causal agent might be *knife*. At both input and output, objects are encoded in terms of perceptual features only (e.g. shape, size, color). The bi-directional connections ensure that the activation will cycle around the network, settling into a stable attractor state. The connection weights are updated according to contrastive Hebbian learning [8].

Each casual agent corresponded to a transformational vector. The transformation vectors encode the relation between the "before" and "after" transformation states of the object. The transformed state of an object (Output 1) is obtained by adding a transformation vector to an object (Input 1). For example, Input 1 (e.g., apple): [0.5 0.2 0.8 0.2 0.4], might be transformed by the vector (e.g., cut): [-0.4 0.0 0.0 0.0.7 0.0] resulting in Output 1 (cut apple): [0.1 0.2 0.8 0.9 0.4].

Input 2 represents a causal agent (e.g. "a knife") which when it occurs alongside certain objects represented at Input 1 (e.g. "an apple") leads to a transformation of that object at Output 1 (e.g. "a cut apple"). Throughout training, the causal agent pattern at output is identical to the causal agent pattern at input.

Training consists in randomly selecting an object and an agent, computing the output states, and updating the weights such that the actual output state produced by the network moves closer to the target output state. One can think of the input and target output states as two temporally contiguous states of the world.

The testing of analogy completion proceeds in a different way from learning of relation information. Priming underlies the network's ability to complete A:B::C:D analogies. Priming occurs because the bi-directional connections allow the hidden and output units to maintain activity resulting from an initial event (e.g., an A:B event). The activity that is maintained in the network impacts on how new input is then subsequently processed (e.g., a C:? input).

Consider the following example. First, the input and output units are clamped with the *apple* at Input 1 and *cut apple* at Output 1 while Input 2 and Output 2 are initially set to 0.5, the resting value. This corresponds to being presented with the information *apple:cut apple* (i.e., the first half of an A:B::C:D analogy). The causal agent is not presented to the network at any point during testing. The network settles into the attractor state it was trained on by filling in Input 2 and Output 2 and arriving at hidden unit activations consistent with the transformation *cutting*. Following this, the pattern corresponding to *bread*, is presented to Input 1 and nothing at Output 1. This corresponds to being presented with the information *bread:?* (i.e., the second probe half of the A:B::C:D analogy). By unclamping the original inputs and outputs and by presenting a different input pattern, the network is no longer in equilibrium and settles into a new attractor. Consequently, the prior priming of the *apple* and *cut apple* transformation biases the network to settle into the attractor state corresponding to *cut bread* at Output 1, appropriately completing the analogy.

The network shows a similar developmental profile to children on a number of measures, capturing well several distinctive phenomena observed with children (see [1] for further details). Despite the good performance of the network, there are a number of concerns, and the subsequent section details an attempt to provide a framework to account for some of these.

2. Extending the model

2.1. *Justification*

There are three main reasons for extending the architecture in [1] to try to produce a more general framework for modeling the development of analogy.

First, since the network uses different layers to represent temporally contiguous states, the earlier account is only capable of learning about domains involving two states (be they of the same object or different objects). This rigidity means that the network is unable to learn about richer environments. For example, in the real world a child might observe an apple being picked up and washed, before being cut, and then eaten. Children can make analogies using any of these different relations, for example, washing, cutting or eating. Consequently, it would be desirable to simulate these multiple states in the network. The most obvious way to achieve this would be to add on extra external layers to represent additional temporally contiguous states. However, there are a number of problems with this approach. The most serious is that we would like to be able to use an open-ended number of states, and this places a large (potentially infinite) computational burden and lack of flexibility on the network. Furthermore, in the real world there is no first time step, corresponding to the first input layer in the network, or a second time step for the second layer. This is particularly true if we want the network to be able to form analogies between semantically distinct domains. A much more efficient and plausible solution would be to represent temporally contiguous states in the world with temporally contiguous states of the same layer in the network.

The second motivation for a more general framework is that the analogies that children make do not only involve target and base domains consisting of two states. For example, 2-year old children are able to make the 3-part analogy based on size between baby, mummy and daddy, and golf ball, football and beach ball. Consequently, it should be possible for the network to form analogies between domains of 3 or more states, in addition to analogies involving 2 states. It is desirable that, in principle, there be no architectural upper limit to the number of states involved in an analogy, although, as in children, there will be de facto practical limits, imposed by computational limitations, such as memory capacity.

The third motivation for extending the account is that most relations used in the world, or at least studied by cognitive scientists, do not involve causal transformations of an object. Consequently, our account needs to be extended to these other relations. However, there is no in principle reason why a similar architecture to that in [1] could not be used with other types of relations: for example, with orderings such as *bigger than*. Similarly, there is no reason in principle why the units in the network have to represent perceptual features only. They could represent semantic features or a mixture of both. The extended framework, described in this chapter, highlights how different relations can be learnt by the network and used to make analogies.

2.2. *A new model*

Figure 3 shows the structure of the more generalized architecture. There are two important differences from the architecture in [1].

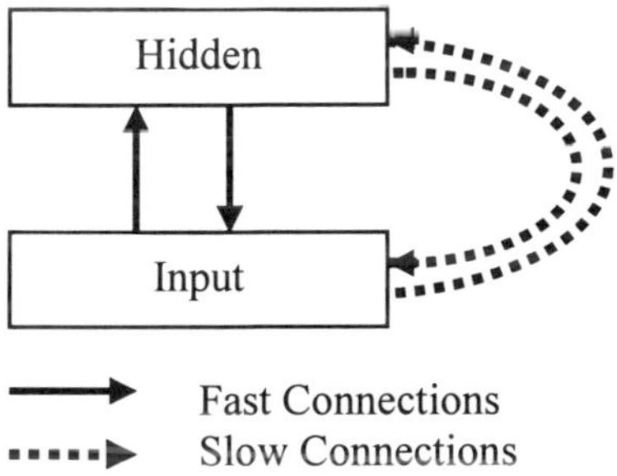

Figure 2. Architecture for the extended framework.

First, instead of three layers there are now only two layers: an internal layer and a hidden layer. These layers are bi-directionally connected with two types of connections: fast and slow connections. The fast connections are identical to the connections in the earlier network (i.e. they pass activation with each cycle of activation). In addition there are slow connections which implement a delay, passing the activation of the network from 20 activation cycles earlier (20 activation cycles gives the network plenty of time to settle into a stable attractor). The hidden layer receives activation from the current state of the input layer as well as the state of the input layer 20 cycles earlier; likewise the external layer receives input from the current state of the hidden layer and the state of the hidden layer 20 cycles earlier. The fast connections have the effect of driving the network into a stable attractor state consistent with the current input before the slow connections kick-in and push it into another attractor basin where the fast weights again drive it into a stable state, now consistent with the current input and the previous stable state of the network. This combination of fast and slow connections working together means that the network can jump from stable state to stable state in an ordered way (see Figure 3).

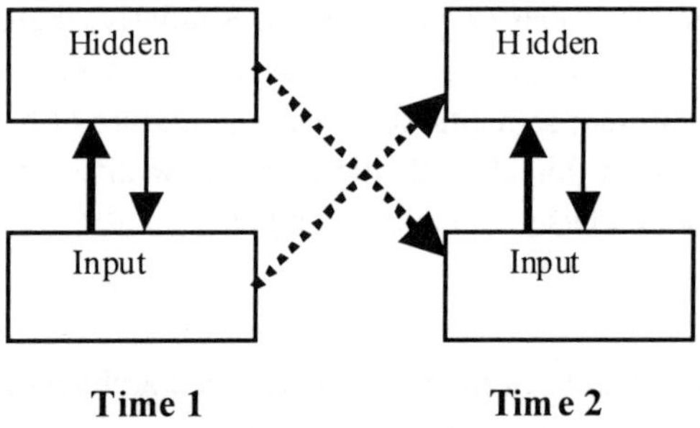

Figure 3. Representation of the architecture over time, with the slow connections pushing the network into a different state at Time 2 from Time 1.

In the pilot network detailed here, the Input layer consists of 20 units and the Hidden layer 40 units. Each unit in each layer links to all the units in the other layer with both fast and slow connections. The net input to each unit is calculated by summing the vector of current input activation multiplied by the relevant fast-weight matrix with the delayed input activation from 20 cycles ago multiplied by the slow-weight matrix. Each unit also receives input from a bias connection. The actual activation of the units is then calculated by applying a sigmoidal activation function to the net input. So, activation values vary between 0 and 1.

2.3. *Training and testing*

As with the original model, the network is again trained with the contrastive Hebbian learning algorithm. The only difference is that the slow connections are not symmetrical. Instead the same delayed activation input value is used for both the +ve and –ve Hebbian terms. The target patterns for the Input layer (i.e. the values the units are clamped to in the +ve phase) are binary vectors. The learning rate is fixed at 0.01 for both the fast and slow connections.

We can see how the new framework operates by looking at how the network solves an analogy involving a causal transformation. In this simple example, the environment is composed of three different objects each of which can be in two states, e.g. *apple* and *cut apple*. The input layer is split into two banks, corresponding to an object and a causal agent. There are two stages to training at each time step, a +ve and a –ve phase. At Time 1, there is a +ve and a –ve phase. In the –ve phase the Input layer is initially clamped to the pattern representing *apple*. Activation then cycles to the Hidden layer via the fast connections. Initially, the slow-connection component of the activation values is calculated by multiplying the slow connections with a resting activation vector of 0.5 for all units. After 5 cycles of activation the Input layers are unclamped and the whole network is allowed to settle into an attractor state. This is done to help pattern completion in the Input layer. Otherwise, the network would not have any way of distinguishing between *apple* and bread at Time 1 and would create a single blended attractor state to represent them both. After 20 cycles, the activation values are stored and the +ve phase starts. The Input layer is clamped to the appropriate pattern, in this case *cut apple*, for all 20 activation cycles and the Hidden layer is then freely updated. Both the fast and the slow weights are then updated with contrastive Hebbian learning [8] using the activation resulting from the +ve and –ve phases. The contrastive Hebbian learning algorithm bases the weight change on the difference between a positive and a negative Hebbian term as follows:

$$\Delta w_{ij} = x_i^+ y_j^+ - x_i^- y_j^-$$

Where x is the post-connection activation in the +ve and -ve activation phases and y is the pre-connection activation in the +ve and -ve activation phases.

At Time 2, in the –ve phase, the Input layer is left unclamped but the activation passing through the slow connections is no longer a resting pattern but the state of activation at Time 1. Over 20 cycles, the Input and Hidden layers are allowed to freely settle into an attractor. In the +ve phase, the Input layer is clamped to the appropriate target pattern, and the Hidden layer is allowed to settle over 20 activation cycles. The fast and the slow weights are then updated. In this way the network learns to move from one state to another in an ordered way.

Testing the network for analogy completion occurs in a similar way to that described in [1] (see Figure 4). The part of the Input layer corresponding to the object, *bread*, is clamped to the appropriate pattern and the network is allowed to settle. After 20 activation cycles the object Input units are clamped to a different pattern, *cut bread*, and the network is allowed to settle. The *cut bread* pattern along with the constraint of the slow connections ensures that the network settles into the attractor state corresponding to *cut bread* and *knife*. The units corresponding to the *knife* pattern then remain active when the object units are clamped to *apple*. Consequently, after 20 activation cycles the network moves from the *apple* state to the *cut apple* state, appropriately completing the analogy.

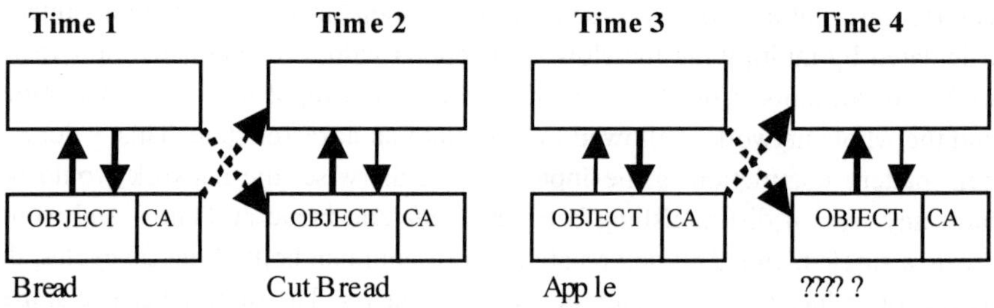

Figure 4. How the network solves an analogy involving causal relations. (CA represents Causal Agent).

3. Conclusions and future directions

Although we have presented a different framework from [1], currently we have only demonstrated that it can solve the same sorts of analogies. To demonstrate the increased flexibility of the new framework, one needs to use it with a three part transitive analogy (see Figure 5). Here, instead of the change in state representing a transformation in an object, it represents an ordering of similar objects, such as *bigger than*. For example, the network moves from an attractor for the *daddy* object, to an attractor for the smaller different object *mummy* ending in the attractor for *baby*.

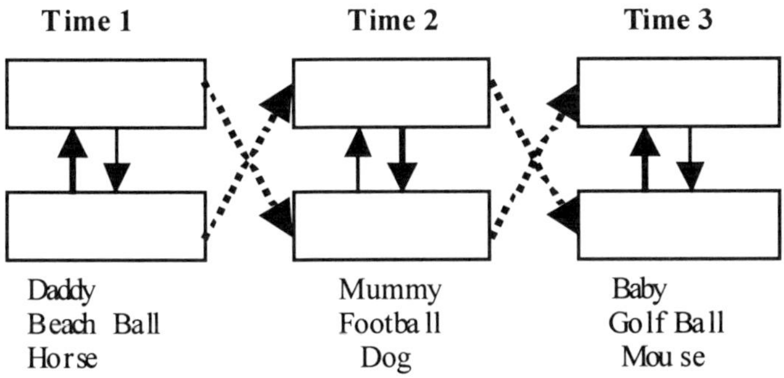

Figure 5. How the network could be applied to transitive analogies.

The idea is that the network would learn different sets of temporally contiguous states, for example, one might be, *daddy, mummy* and *baby,* another might be *beach ball, football* and *golf ball.* The network could solve transitive analogies in the same way as it solved the two part analogies above. First one set of temporally contiguous objects is presented to the network, e.g. *daddy/mummy/baby*. Subsequently the network is presented with an object from another set of objects, e.g. *beach ball.* If biggest person and biggest ball have shared features then the slow connections will push the network into the attractor state corresponding to the smallest *ball*, i.e. *golf ball*, appropriately completing the analogy.

In summary, this chapter has presented a framework for extending [1] by including temporal attractors to be more flexible and account for the development of a greater variety of analogies. We have replicated the analogy completion of the original model with the new account. Furthermore, we have suggested how the new framework could be used to model the development of a substantially different type of analogy

Acknowledgements

The work was supported by an ESRC (UK) studentship award to Robert Leech. Additional support came from ESRC grant R000239112 and EC RTN grant HPRN-CT-2000-00065.

References

1. Leech, R., Mareschal, D., & Cooper, R. P. (2003). A connectionist account of analogical development. *Proceedings of the 25th Annual Conference of the Cognitive Science Society.*

2. French, R. M. (2002). The Computational Modeling of Analogy-Making. *Trends in Cognitive Sciences, 6*, 200-205.
3. Goswami, U. (1996). Analogical reasoning and cognitive development. *Advances in Child Development and Behaviour, 26*, 91-138.
4. Chen, Z., Sanchez, R.P., & Campbell, T. (1997). From beyond to within their grasp. *Developmental Psychology, 33*, 790-801.
5. Gentner, D., & Toupin, C. (1986). Systematicity and surface similarity in the development of analogy. *Cognitive Science, 10*, 277-300.
6. Goswami, U., & Brown, A.L. (1989). Melting chocolate and melting snowmen: Analogical reasonsing and causal relations. *Cognition, 35*, 69-95.
7. Rattermann, M. J., & Gentner, D. (1998). More evidence for a relational shift in the development of analogy: Children's performance on a causal-mapping task. *Cognitive Development, 13*, 453-478.
8. O'Reilly, R. C. (1996). Biologically plausible error-driven learning using local activation differences. *Neural Computation, 8*, 895-938.
9. Larkey L., & Love, B. (2003). CAB: Connectionist Analogy Builder, *Cognitive Science*, *27*, 781-794.

ON THE EVOLUTION OF IRRATIONAL BEHAVIOUR

JOHN A. BULLINARIA
School of Computer Science, The University of Birmingham
Edgbaston, Birmingham, B15 2TT, UK
j.bullinaria@physics.org

Many aspects of human and animal behaviour require individuals to learn quickly how to classify the patterns they encounter. One might imagine that evolution by natural selection would result in neural systems emerging that are very good at learning things like this. Explicit simulations of the evolution of simple developmental neural systems confirm that such rational behaviour can indeed emerge quite easily. However, the same simulations also reveal that there are situations in which evolution seems to let the species down, and populations emerge that appear to perform rather irrationally. There are actually many reasons why this might happen. I shall present the results from a selection of my simulations that begin to explore the issues involved.

1. Introduction

A standard definition of *irrational behaviour* is the performance in a manner which goes against one's objectives [10]. This can range from minor but persistent miscalculation of probabilities resulting in poor decisions, through to apparently totally random actions. Understanding why individuals should ever act in such a way is an interesting psychological research area in its own right [5], and there are many other fields in which irrational behaviour is studied in detail, e.g. finance [7] and law [8]. In this paper I shall begin to explore how such behaviour might evolve in simple neural network systems.

Many aspects of human and animal behaviour require individuals to learn quickly how to classify the patterns they encounter, and how to act on those classifications. For example, which foods are good or safe to eat, which other animals should be feared, which environments should be avoided, when to sell your stocks and shares, and so on. One might imagine that evolution by natural selection would result in neural systems emerging that are very good at learning things like this. This would certainly fit in well with modern ideas of evolutionary psychology as an explanation of human behaviour [6]. The question that still needs addressing is: why do individuals so often apparently get things wrong and act in an irrational manner (e.g. losing money by selling their shares too early [7], or by breaching a contract [8])? If we can understand such behaviour, and perhaps predict that behaviour, then, depending on our own motives, we can maybe help, or take advantage of, that behaviour.

Over the past few years I have been performing increasingly biologically realistic simulations of the evolution of neural systems with view to better understanding how various aspects of the human information processing system works. These have included the evolution of systems that perform adaptive control [4] and general binary mappings [3], and also explorations into the evolution of modularity in these systems [2]. Here I shall concentrate on simulations of the evolution of simple developmental neural systems that are required to classify various types of sensory information. We shall see that rational behaviour, in the form of learning to classify quickly and reliably, can emerge quite easily. However, in the same simulations we often find that there are situations in which evolution seems to lets the species down, and populations emerge that appear to perform rather irrationally. There are actually many reasons why this might happen, and often the seemingly irrational behaviour does actually make good sense. One can then attempt to discover which aspects of human behaviour this might correspond to [5, 7, 8, 10]. In the remainder of this paper I shall present the results from a small selection of my simulations that begin to explore the issues involved.

2. The Evolutionary Models

The field of Evolutionary Neural Networks is already quite mature [11], but here we are interested in simulations that are more biologically realistic than is usually the case. Two aspects are particularly important:

1. *The Evolutionary Population.* Biological populations usually contain individuals of all ages, who learn from their environment, compete amongst themselves for the opportunity to live and procreate, but eventually die of old age. We want to avoid a traditional generational approach [11] in which whole populations are created, taught, and tested one generation at a time.
2. *The Environment.* Biological individuals learn by sampling a continuous range of environmental conditions, and must generalize to cope with sensory information they have never seen before. We want to avoid the traditional training data split into fixed training and testing sets [1].

Naturally, we shall still have to make an enormous number of simplifications and abstractions, but an effort will be made to accommodate these two aspects.

We shall work with a population of individual neural networks, each specified by a number of 'innate' parameters that are recorded in its genotype. For simplicity, we restrict ourselves to simple feed-forward networks with one hidden layer trained by gradient descent with momentum [1]. The training data

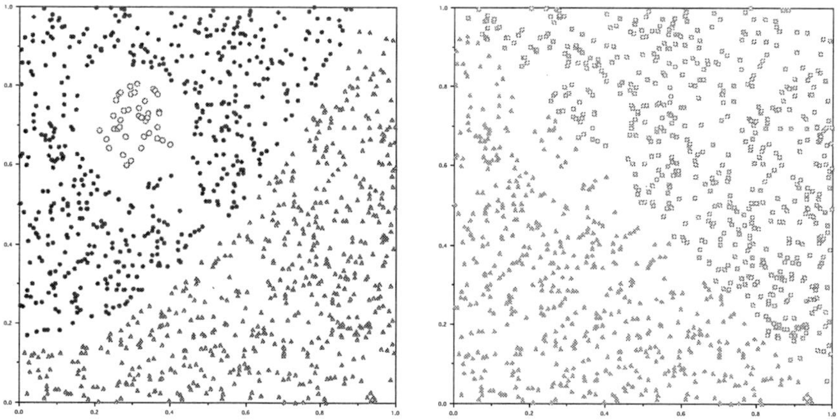

Figure 1: Samples from the dual classification training data distributions. One case (left) has three classes (two large separated classes and one small surrounded class), while the other (right) has two classes (two different large separated classes).

will specify the number of inputs and outputs. The genotype will specify the network architecture, the initial weight distributions, and the learning rates. During each simulated year, the least fit individuals and the oldest individuals will die (i.e. be removed from the population) to be replaced by the children of the fittest individuals. The genotype of each child will depend on the genotypes of its two parents, plus random mutations. We can then expect evolution by natural selection to produce increasingly fit populations.

3. Simulation Details

Here it is convenient to consider a particularly simple environment in which each individual has just two continuous valued sensor inputs, and must perform two distinct classifications on them. Figure 1 shows samples from the distributions used in all the simulations. The inputs might, for example, correspond to particular components of smell and colour, and the two output cases could be classes of usefulness, such as {good tasting, bad tasting, poisonous} and {easy to get, hard to get}. Alternatively, the inputs could be financial indicators, and the outputs could be trading strategies.

During each individual's life, they will experience a stream of samples from the input distribution, and must learn quickly how to perform well on any new samples they come across. The individual neural networks need to perform two distinct classifications based on data in a single input space. We shall therefore

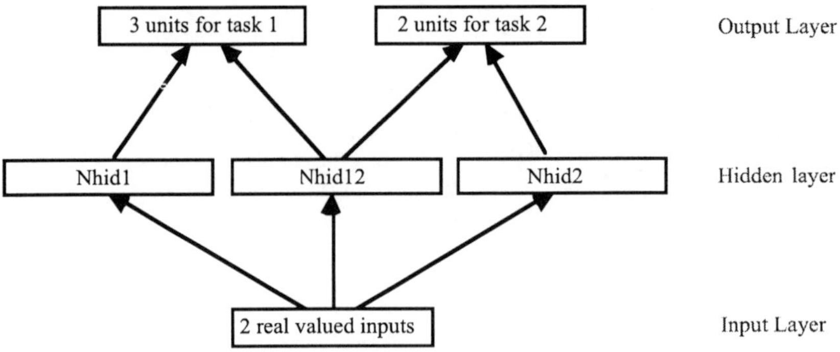

Figure 2: Architecture of the individual neural networks.

parameterize the networks in such a way that they may evolve either a modular, or non-modular, architecture to do this [2]. At the same time, they are also allowed to evolve their own associated learning algorithm and parameters [3].

The basic evolutionary implementation has been described in some detail previously [2, 3, 4], so I shall simply summarize the main details here. Each individual neural network will have the same basic architecture as shown in Figure 2, with the same total number of hidden units *Nhid* = 36. Then two parameters *Con1* = *Nhid1* + *Nhid12* and *Con2* = *Nhid12* + *Nhid2* are sufficient to specify how many hidden units connect to each output block, and how many connect to both. If *Nhid12* takes on a zero value, we have a totally modular architecture, and if both *Nhid1* and *Nhid2* become zero, we have a totally non-modular architecture [2]. Each new network is created with random initial weights drawn from innately specified distributions, and learns using a learning algorithm with innately specified parameters [3]. For each network it proves appropriate to have four uniform initial random weight distributions $[-iw_L, +iw_L]$, four learning rates η_L, and a momentum parameter α, in which L refers to the network layer (input to hidden, hidden biases, hidden to output, output biases). It is likely that the optimal parameters will depend on which gradient descent cost function is used. Here we shall compare using a sum squared error cost function with targets of 0.1 and 0.9 as used in the original modularity study of Rueckl et al. [9], and the cross-entropy cost function that is now known to be better for classification problems [1, 3].

The evolutionary survival/procreation fitness is inversely and logarithmically related to the total number of classification errors during the last simulated year as 1/log(1+*ErrorCount*). Each individual experiences 125 randomly selected data samples per simulated week. The population size is fixed at 100, with

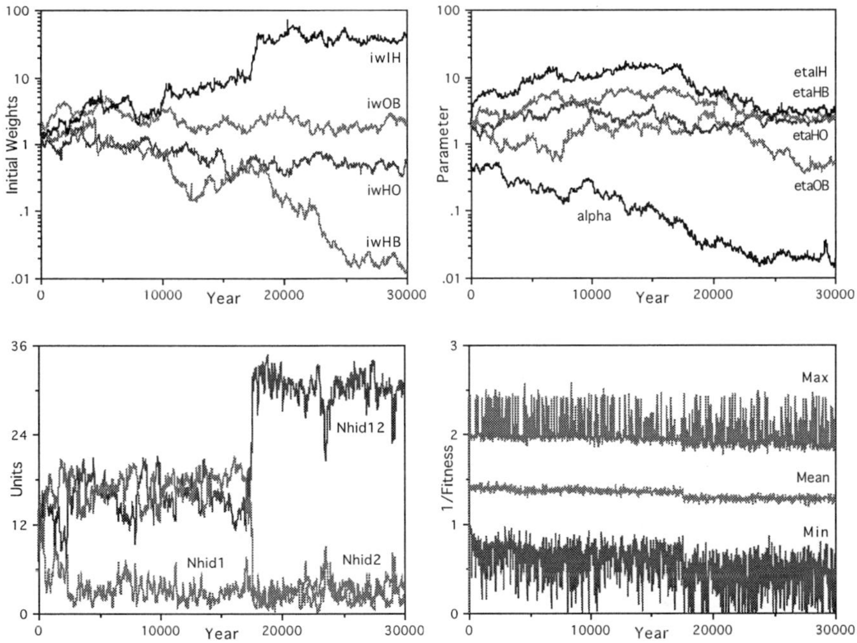

Figure 3: Typical evolution of the mean initial weight distributions, learning rates, architecture parameters, and population fitnesses for the sum squared error model.

death rates chosen to give realistic age distributions, with life expectancies of about thirty years. The procreation and mutation parameters are chosen to maintain genetic diversity without introducing too much noise. The resulting simulations are rather course, but good enough to provide reliable results.

4. Simulation Results

The complexity of our tasks were chosen so that our simple neural networks could evolve to perform essentially perfectly on them. If populations emerge that perform in a manner which does not achieve their objective, then this can be classed as an example of irrational behaviour, and we will wish to investigate further, and perhaps relate it to corresponding patterns of human behaviour.

Long simulation runs indicate that the evolving populations usually stabilize by about 30,000 simulated years. Figure 3 shows the typical evolutionary pattern when the sum squared error cost function is used for learning. We find that similar values emerge for the four learning rate parameters, but there are very large variations between the evolved initial weight

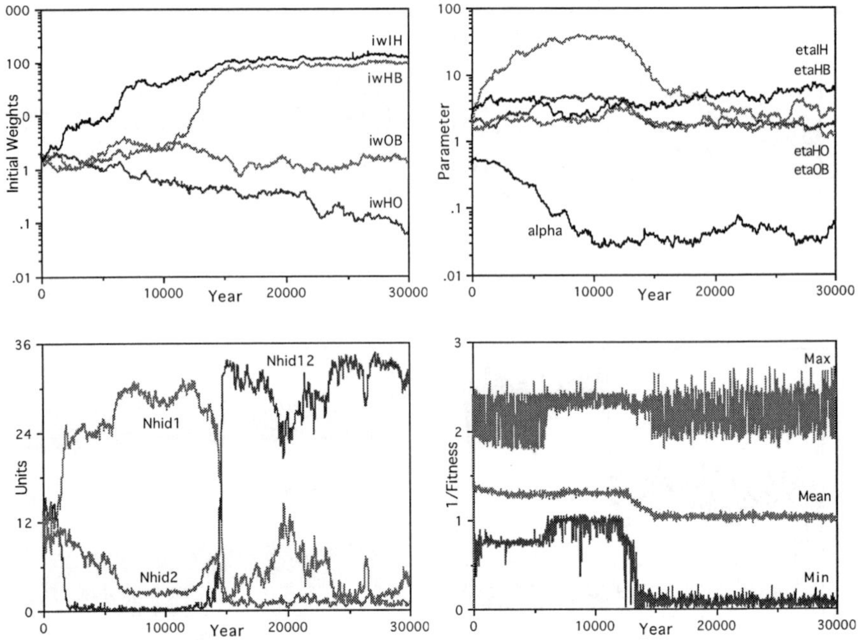

Figure 4: An unusual evolutionary pattern for the mean initial weight distributions, learning rates, architecture parameters, and population fitnesses, for the same underlying system as that of Figure 3.

distributions. We see that a strange equilibrium state exists for about 18,000 years, before the rapid emergence of a totally non-modular architecture and a step change in the mean fitness level of the population. Notice that even the best performing individuals rarely manage to learn the task to perfection.

As always, it is important to test the robustness of the simulation results. Numerous runs of the same system, with different random initial conditions and training data, reveal that the populations can end up in at least two different local maxima of evolutionary fitness. Figure 4 shows an evolutionary run that ends up with a final population that has significantly better mean fitness and best individual fitnesses than the typical run shown in Figure 3. The final non-modular architecture and learning rates are much the same as before, but the initial weight distributions are very different, with a much closer correspondence here between the input-hidden initial weights and the initial hidden unit biases. Interestingly, between years 6,000 and 12,000 the populations go through a pure modular phase in which the best individual performances are worse than before, but without making the mean fitness worse. The individuals there have evolved

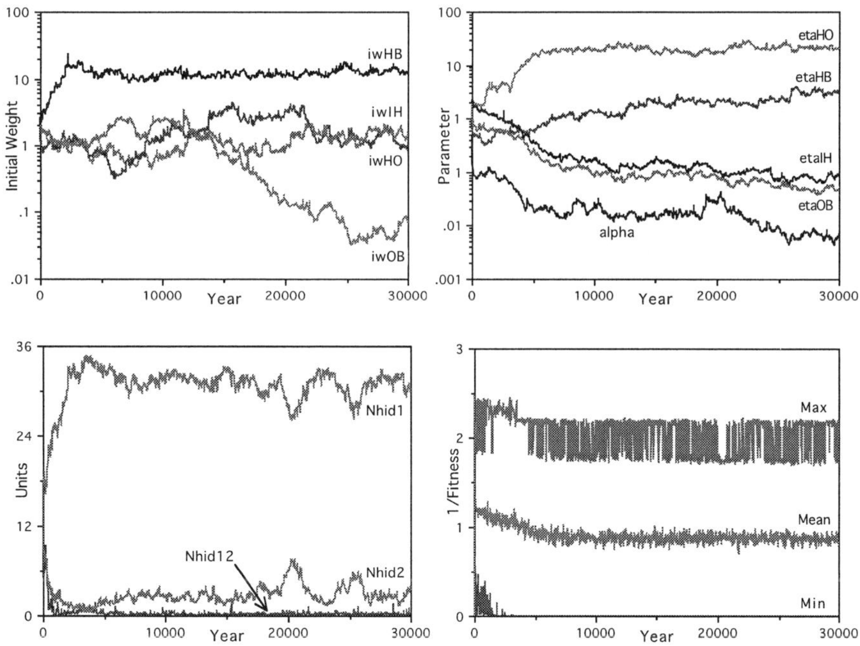

Figure 5: Typical evolution of the mean initial weight distributions, learning rates, architecture parameters, and population fitnesses for the cross-entropy model.

to learn faster early on, at the expense of their final performances. At the end of this phase the architecture and initial hidden unit biases drift into the better non-modular regime, and the learning rates soon follow suit, ending up with a population of individuals that perform better throughout their lives.

My earlier study on modularity evolution [2] indicated a strong dependence on the gradient descent learning cost function, so the above simulations were repeated using the cross-entropy cost function [1, 3]. Figure 5 shows the patterns of evolution typically observed in this case. As one might expect, the evolved initial weight distributions and learning rates are rather different to the sum-squared error case. Moreover, the population mean fitnesses are better, and individuals regularly learn the tasks to perfection. The individuals achieve this by evolving a modular architecture, with the numbers of hidden units assigned to each of the two tasks in agreement with their relative difficulty, and in line with the modular phase observed in Figure 4.

In my previous study of modularity using the simpler fixed "What-Where" training data [2], it was the sum squared error systems that evolved modular architectures, and the better performing cross-entropy systems evolved non-

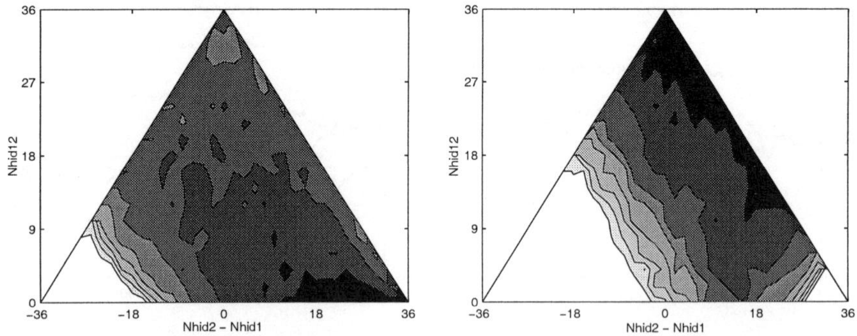

Figure 6: Mean learning times for the cross-entropy model as a function of architecture using the evolved learning parameters (left) and parameters evolved for a fixed non-modular architecture (right). Darker shading means faster learning.

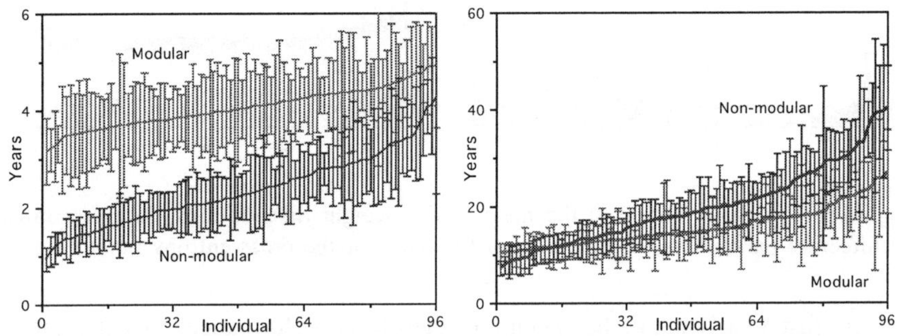

Figure 7: Ages at first full week of perfect performance (left) and first full year of perfect performance (right) for the modular and non-modular individuals.

modular architectures. We can investigate the fitness dependence on architecture by training a series of networks using the evolved initial weight distributions and learning rates, and plotting as a function of architecture the mean times taken to reach the first full week of perfect performance. This is shown for the cross-entropy case on the left of Figure 6. The fastest learners do indeed occur in the modular regime (*Nhid12* = 0) with more hidden units dedicated to the harder task (*Nhid2* – *Nhid1* > 0) as the evolutionary simulations suggest.

The question remains as to what architecture comes out best if we were to optimize the learning parameters for a totally non-modular architecture? It is easy to fix the architecture and evolve those parameters, and then test the learning times again. This is shown on the right of Figure 6. Now the fastest learners are in the non-modular (*Nhid12* = 36, *Nhid1* = *Nhid2* = 0) regime.

Moreover, if we run a full evolutionary simulation with evolvable architectures, starting with the non-modular population, the population stays non-modular, indicating that it is a stable local maximum of evolutionary fitness.

The natural question to ask is: does the non-modular population perform better than the evolved population? Individuals in both populations regularly learn the task to perfection, but it is the non-modular population that has the lower mean fitness, indicating that its individuals tend to learn faster. We can explore this more carefully by plotting the learning times from Figure 6 for each individual. The left graph of Figure 7 shows the means and standard deviations over ten sets of training data, with the individuals ordered for clarity. We see that the non-modular individuals are indeed better, which makes us wonder why the modular population evolved. The crucial fact is that our simulated evolution depends on the performance over a full year, not a single week. The right graph in Figure 7 shows the times taken to reach the first full year of perfect performance, and now the modular population (with mean 15.7) does out-perform the non-modular population (mean 20.1). This factor also provides a likely explanation of why using fixed training sets, rather than the more realistic continuous data sampling used here, led to different architectures evolving [2].

5. Discussion and Conclusions

I have outlined a framework for simulating the evolution of neural network systems that must learn to perform simple classification tasks, and presented a small selection of my simulations which show how behaviour can evolve which is less than optimal and might therefore be classed *irrational*. By making these simulations sufficiently realistic, one can hope to come to a better understanding of the emergence of irrational behaviour in real systems.

We have seen examples of several factors that can lead to what looks like irrational behaviour, but make sense when we investigate why they arise. Often it is simply the result of conflicting requirements on the system. We found:

1. Reliable good performance might require slower learning, and may even be more important than good average performance,
2. Fast initial learning might mean bad final performance, and good final performance might need long learning times

In the psychology and economics literature, the term *bounded rationality* is often used to describe the results of such constrained optimization [5, 8]. Our simulations have also found purely evolutionary factors:

3. Optimal behaviour may simply be difficult or extremely slow to evolve,

4. Certain non-evolvable aspects of the learning mechanism can place limitations on what objectives are actually achievable,
5. Accidents of evolutionary history may leave populations in local maxima of fitness, even though they could in principle perform better.

Further simulations are likely to reveal other important factors, such as:

6. Robustness against mutations might limit performance levels,
7. The importance of co-evolution and coping with very noisy data.

It is clear that there is much scope for further research in this area. Perhaps this paper will give others an idea of the range of effects that can come out of simple evolutionary neural network simulations.

References

1. Bishop, C. M. (1995). *Neural Networks for Pattern Recognition*. Oxford, UK: Oxford University Press.
2. Bullinaria, J. A. (2001). Simulating the Evolution of Modular Neural Systems. In *Proceedings of the Twenty-third Annual Conference of the Cognitive Science Society,* 146-151. Mahwah, NJ: Lawrence Erlbaum.
3. Bullinaria, J. A. (2003). Evolving Efficient Learning Algorithms for Binary Mappings. *Neural Networks,* **16**, 793-800.
4. Bullinaria, J. A. (2003). From Biological Models to the Evolution of Robot Control Systems. *Philosophical Transactions of the Royal Society of London* A, **361,** 2145-2164.
5. Chase, V. M., Hertwig, R. & Gigerenzer, G. (1998). Visions of Rationality. *Trends in Cognitive Science*, **2**, 206-214.
6. Mitchell, M. (1999). Can Evolution Explain How the Mind Works? A Review of the Evolutionary Psychology Debate. *Complexity,* **4**, 17-24.
7. Montier, J. (2002). *Behavioural Finance: Insights into Irrational Minds and Markets.* New York, NY: Wiley.
8. Parisi, F. & Smith, V. L. (2003). *The Law and Economics of Irrational Behavior*. Chicago, IL: Chicago University Press.
9. Rueckl, J. G., Cave, K. R. & Kosslyn, S. M. (1989). Why are "What,, and "Where,, Processed by Separate Cortical Visual Systems? A Computational Investigation. *Journal of Cognitive Neuroscience*, **1**, 171-186.
10. Tribus, M. (1969). *Rational Descriptions, Decisions and Designs*. New York, NY: Pergamon Press.
11. Yao, X. (1999). Evolving Artificial Neural Networks. *Proceedings of the IEEE,* **87**, 1423-1447.

MULTIPLE PERSON INFERENCES: A VIEW OF A CONNECTIONIST INTEGRATION

FRANK VAN OVERWALLE
Vrije Universiteit Brussel.
Pleinlaan 2,
B-1050 Brussel, Belgium
E-mail: Frank.VanOverwalle@vub.ac.be

This paper provides a connectionist account of the processes underlying the multiple inference model of person impression formation proposed by Reeder, Kumar, Hesson-McInnis and Trafimow [7]. First, in a replication and extension of one of their main studies, I found evidence for discounting of trait inferences when facilitating situational forces were present consistent with earlier causality-based theories, while at the same time I replicated the lack of discounting in moral inferences as documented and predicted by Reeder et al. [7]. Second, to provide an account of how these different and sometimes contradictory inferences are formed and integrated in a coherent person impression, I present a recurrent network model that automatically integrates these inferences, resulting in a pattern that closely reproduces the observed data.

1. Introduction

Although person impression research is replete with studies on how context and behavior interact to determine one's impression about a person, in the past, little attention has been given to other inferences that people may make in the process. In a recent study, Reeder et al. [7] proposed a *multiple inference model* to account for the many inferences that are made and used to arrive at a person impression. They documented that besides trait and causal inferences, perceivers routinely infer much more about the actor, such as his or her motives for engaging in a given behavior and its social implications. Although it is widely accepted that perceivers take note of an actor's goals, Reeder et al.'s important insight is that these inferences impact on the impression formation process. For instance, when an actor engages in aggressive behavior, we may infer not only to what extent he or she is an aggressive person, but also which reasons or motives behind the aggressive act may have compelled the actor to behave the way he or she did. By taking this information into account, Reeder et al. argued, we may end up with a completely different impression about the person. This is interesting, because it can potentially explain why people sometimes fall prey to attribution biases, like the fundamental attribution error. This bias refers to the well-known phenomenon that in explaining an actor's behavior, observers often do not sufficiently take into account the facilitative contextual forces and fail to use this information to discount the contribution of the actor. Perhaps, motives may explain in part why this bias occurs.

The aim of the present paper is to replicate and extent the crucial finding of Reeder et al. [7] that perceivers apparently fail to apply discounting when inferring a behavior-implying trait due to additional inferred motives. In addition, because Reeder et al. [7] did not provided a formalization of how multiple inferences are derived, I propose a connectionist account of this process. This is part of an ongoing attempt to provide more precise computational implementations of diverse phenomena and processes in social cognition. Inspired by the ever-increasing success of connectionist models in cognitive psychology, a number of social psychologists developed connectionist models of causal attribution [6, 11], cognitive dissonance [9, 12], group impression formation and change [3, 14] and person impression formation [2, 10, 13].

2. Multiple Inferences

To illustrate the multiple inference perspective, consider the following scenario used in Reeder et al's [7] research: John, a participant in a psychological experiment, is told that he has the choice to either reward another research participant or punish the other participant, by giving an extra dollar or delivering a shock. John is further told that the other participant is facing a similar decision about whether he would reward or shock John, and that the other participant would be the first to choose. In one situation, the other participant decided to deliver the shock, and thus encouraged an aggressive response from John. In the other situation, the other participant decided to reward John, and thereby discouraged an aggressive response from John. In both situations, however, John decided to deliver a shock to the other participant. In which situation would we consider John the most moral or the most selfish character?

Reeder et al's [7] hypothesis was that inferences about morality are based on the perceived motives of the target. In the above aggression encouraging scenario where the other participant punished first, the perceived motive may include revenge, so that John's aggression in response to this sort of provocation seems quite natural and legitimate. Reeder et al. [7] argued that "motives of this sort may be viewed as relatively positive (or less negative)" and "it follows, therefore, that perceivers' inferences about morality should be more positive" (p. 792). In contrast, other aggressive situations may revolve around a selfish desire for profit. For instance, image the same scenario but now no mention is made of the other participant's choice, but rather that the experimenter either encouraged aggression by offering John a $5 incentive for delivering the shock or discouraged aggression by offering the same incentive for rewarding the other participant. In this aggression encouraging situation, Reeder et al. [7] argued, "the potential motives underlying instrumental aggression converge on selfish desires for reward. Perceivers may react negatively to such motives" (p. 792).

Consistent with their hypothesis, across four experiments, Reeder at al. [7] found a strong decrease of morality when the aggressor acted out of selfishness as opposed to revenge. Thus, when John reacted to the other participant's punishment that encouraged aggressiveness (*reactivity condition*), he was seen as more moral and less selfish than when the experimenter provided financial incentives that encouraged aggression (*instrumentality condition*). These findings indicate that the motives underlying someone's behaviors seem to contribute strongly to inferences of morality. Hence, Reeder et al. [7] concluded that "inferences about the morality of an aggressor are based more on the perceived motives of the target than on the presence of facilitating situational forces" (p. 789).

3. Discounting and Motives

An important implication of Reeder et al.'s [7] findings is that they challenge general causality-based models of dispositional inference such as those of Kelley [4] and Gilbert [1], in that perceivers apparently failed to apply Kelley's discounting principle. According to the discounting principle, perceivers consider whether the behavior appears to have been caused by the actor's disposition or by situational forces, and assume that both causes operate in a hydraulic fashion. Hence, when the situation encouraged aggression, perceivers should subtract out the effect of the situation, with the result that only a minimal amount of aggression is attributed to the person. Conversely, when the situation discourages aggression, perceivers should not subtract out the effect of the situation, and attribute a great amount of aggression to the person.

In contrast to these assumptions, the results of Reeder et al. showed that although perceivers discounted selfishness in the revenge (reactive aggression) condition, they did not discount selfishness when they were paid for the aggressive behavior (instrumental aggression condition). Reeder et al. argued that this was so because the motive of revenge was legitimate (and not so negative), whereas the motive of selfish gain was not legitimate and is consistent with low morality. Hence they concluded that the "perceiver's use of the discounting principle depended on the type of situational force that was operating" (p. 799).

4. Measuring Traits and Motives: A Replication

Although it is possible that discounting is not applied when making morality judgments, the crucial question is whether perceivers also fail to apply discounting when inferring the correspondent (i.e., behavior-implying) trait? A limitation of most studies by Reeder et al. [7] is that they did not measure inferences about the correspondent trait aggressiveness. This makes it difficult to compare traits against motives during the same inference process. While a correspondent trait aggressiveness is very close to the actor's aggressive behavior in semantic terms, inferred motives and morality may have greater conceptual breadth and may be more distant to the aggressive behavior. This leaves open the possibility that situational forces are not immediately subtracted out of a moral judgment, as they constitute an essential part in the evaluation and definition of a moral act. Thus, while situational discounting may be absent in judgments of motives and morality, in attributing traits of a person, some discounting of facilitating situational pressures may still take place. In order to verify this hypothesis, I replicated the second study of Reeder et al. [7] and extended it with a correspondent trait measure.

4.1. *Method*

Participants were 88 male and female students from the Vrije Universiteit Brussel, Belgium. I used the same materials (translated to Dutch) from Study 2 by Reeder et al. [7] which depicted John's reactions in the psychology experiment as described earlier in the introduction. To recall, in the *reactivity scenario*, John received a shock from another participant (aggression encouraged) or did not received a shock from another participant (aggression discouraged). The *instrumentality scenario* was similar, except that no mention was made of the other participant's choice. Rather, the experimenter said that he would financially reward John for shocking the other participant (aggression encouraged) or for not shocking the participant (aggression discouraged). All conditions ended by John shocking the other participant.

Immediately following the description of the situation and behavior, the participants rated John on several scales taken from Study 2, except when noted otherwise. Trait aggressiveness was measured with the following item (from Study 4): "How aggressive is John in general, in his everyday interactions" (1 = *not aggressive*; 10 = *very aggressive*). Morality was measured by an item dealing with John's selfishness: "How selfish is John (1 = *not at all selfish*; 10 = *very selfish*) and an item dealing with morality "How moral is John" (1 = *not at all moral*; 10 = *very moral*). In addition, motives were measured by an item dealing with John's motivation to earn money: "To what extent is John motivated to earn money" (1 = *not at all motivated*; 10 = *very motivated*), and John's motivation to revenge himself by an item (from the coding of the open responses in Study 2): "To what extent is John motivated to revenge himself" (1 = *not at all motivated*; 10 = *very motivated*). Two items served as manipulation checks and asked participants to rate the strength of the situational forces related to provocation: "Did the other participant do anything harmful or inappropriate to John?" (1 = *not at all*; 10 = *very much*), and related to reward: "Could John gain something by hurting the other player? (1 = *not al all*; 10 = *very much*).

4.2. *Results*

The results showed that the participants discounted *trait aggressiveness* in situations that encouraged aggression in line with causality-based theories, whereas their *moral judgments* were determined by personal motives as Reeder et al. [7] predicted (see Table 1). An analysis of variance on *trait aggressiveness* revealed only a main effect of situation, $F(1, 84) = 17.06$, $p < .0001$. The results showed lower ratings of aggressiveness when the situation encouraged aggression than when the situation discouraged aggression regardless of type of scenario.

	Reactivity		Instrumentality	
	Aggression encouraged	Aggression discouraged	Aggression encouraged	Aggression discouraged
Trait Aggressiveness	4.78_a	6.32_b	4.95_a	6.45_b
Morality	4.77_a	4.09_a	4.05_a	3.68_a
Selfishness	5.59_b	5.73_b	6.82_c	4.18_a
Motivation to Earn	5.05_b	4.41_b	7.91_b	2.05_a
Motivation to Revenge	7.50_c	4.41_a	4.50_a	6.00_b

Table 1. Perceptions of an Aggressive Target Person as a function of Scenario and Situation (Replication Study). All cells N = 22, except for Motive to Earn which contained one missing response. Means in a row with a different subscript differ significantly from each other (Newman-Keuls tests with p < .05)

In contrast, replicating Study 2 of Reeder et al. [7], the predicted interaction between type of scenario and situation was found for ratings of *selfishness*, *motivation to revenge* and *motivation to earn money*, $Fs(1, 84) = 9.49—49.06$, $ps < .01$. No effects were found for ratings of *morality*. For the reactivity scenario, Newman-Keuls tests showed that the motivation to revenge was stronger when the situation encouraged aggression (when the other participant shocked John first) than when the situation discouraged aggression (when the other participant did not John shock), $p < .001$. Conversely, in the instrumentality scenario, tests showed that ratings of selfishness and motivation to earn money were stronger, and that the motivation to revenge was weaker when the situation encouraged aggression (when the experimenter rewarded shocking the other participant) than when the situation discouraged aggression (when the experimenter rewarded not shocking), *ps* < .01.

Taken together, the results of the aggressiveness ratings are consistent with causality-based theories. We saw discounting of aggressiveness ratings whenever external situational forces encouraged aggressive behavior, irrespective of the type of scenario. In contrast, the selfishness ratings are consistent with Reeder et al.'s [7] suggestion that participants tend to infer underlying motives and morality that are very different in the two scenarios. When the situation encourages aggression, the actor's aggressive behavior is mainly perceived as being driven by a legitimate motive to revenge in the reactivity scenario, and by a selfish motive to earn money in the instrumentality scenario. The results did not replicate Reeder et al. [7] morality ratings, presumably because this type of abstract inference is somewhat more culturally dependent and difficult to translate properly with its original (American) meaning preserved.

5. Simulation with a Recurrent Network

The results of the replication study showed that discounting took place for aggressiveness trait inferences in all conditions that encouraged aggression, but not for moral inferences. Thus, inferred traits did not fall prey to the fundamental attribution error, whereas inferred morality did. How can these divergent inferences be explained? Reeder et al. [7] proposed that an inferred motive tends to be evaluated either positively (or justified) or negatively (or unjustified), and that these evaluative reactions are reconciled with their trait inferences. Thus, inferences of motives and traits are based on their evaluative consistency. However, the authors did not specify the process by which this integration takes place.

The aim of this section is to demonstrate how a connectionist approach may provide an answer to the seeming paradox of divergent motives and traits. I applied the auto-associator network with the delta learning algorithm [5] used also by [10] and [13] in their simulations of classic findings in person impression formation. My replication study of [7] was modeled using a network architecture consisting of an actor node connected to a trait-implying behavior node (i.e., shock), and additional nodes that reflected situational forces or affordances, including the opportunity to gain money and the other participant (see Figure 1). I also assume that perceivers implicitly categorize the behaviors and motives as socially good (prosocial) or bad (asocial). Although these implicit categorizations are not essential for the present simulation as they are all asocial, they were included because Reeder proposed evaluative consistency as an essential part of the integration of trait and moral inferences. Simulations of more recent studies by Reeder et al. [8], not reported here, show more compellingly that person inferences may depend on these implicit evaluative categorizations.

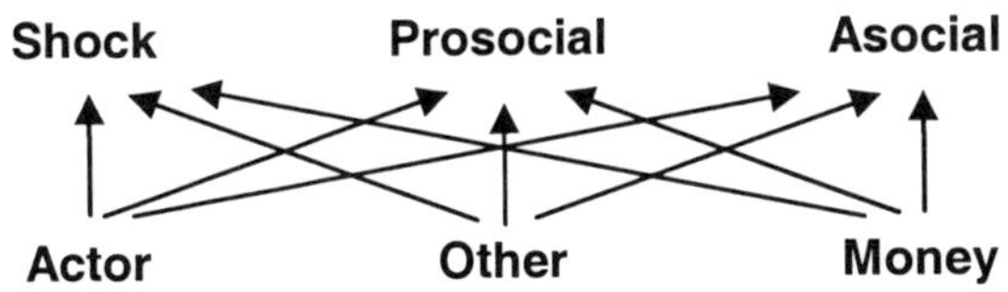

Figure 1. Architecture of the network. Other = Other Participant. Only the most important upward connections are drawn, but all downward and all lateral (in two directions) connections were also included in the simulation.

5.1. *Method*

I simulated each experimental condition separately. To reflect the idea that perceivers take some time to analyze the information provided, in each condition, I provided two learning trials. Table 2 lists the hypothesized learning history (see also [13]).

During learning, I used learning rate = 0.25 with linear activation updating and number of internal cycles = 1. The external activation levels listed in Table 2 received additional random noise generated from a Normal distribution with mean 0 and standard deviation 0.20. The starting values of the weights were set at 0.15. This reflects the idea that perceivers typically start with the assumption that the actor possesses the inferred traits and motives to some minor degree.

At the end of each simulated experimental condition, to simulate the empirical dependent measures, test trials were run by prompting certain nodes of interest (i.e., turning on their activation), and the resulting output activation in other nodes was recorded. For instance, to test trait inferences, the actor node was turned on and the resulting activation of the trait-implying behavior node (without any additional external activation) was read off. Similar test procedures for the other dependent variables can be seen from the bottom panel of Table 2.

The simulations were repeated for 100 "participants", and the results were then averaged. These simulations were then verified by comparing the resulting mean test activations with observed experimental data. Given that the resulting activation values and experimental results are difficult to compare quantitatively, I examined only the general pattern of activations and projected them visually onto the observed data (i.e., I re-scaled the obtained mean test activations by linear regression with a positive slope).

Table 2: Hypothetical Learning History and Measures of Inferences as a function of Scenario and Situation

	Nodes					
	Actor	Shock	Other	Money	Prosocial	Asocial
	Conditions					
Reactivity / Encouraged	1	1	1	0	0	1
Reactivity / Discouraged	1	1	0	0	0	1
Instrument. / Encouraged	1	1	0	1	0	1
Instrument. / Discouraged	1	1	0	-1	0	1
	Measures					
Aggression	1	?	0	0	-?	?
Selfishness	1	0	0	?	-?	?
Motive to Gain	1	0	0	?	0	0
Motive to Revenge	1	0	?	0	0	0

Note. Representation of the simulation based on the experimental design of [7, Study 2]; Other = Other Participant; Money = Gain of Money. Cell entries denote external activation to which noise was added that was randomly drawn for each trial from a Normal distribution with mean 0 and standard deviation of .20. Starting values of the weights were 0.15. The simulation was run separately for each condition, and in each condition each trial was repeated twice. After each condition, the "Measures" section was run.

5.2. *Results*

As can be seen from Figure 2, the simulated values reproduced the observed ratings very closely. The aggressiveness traits failed to show a significant interaction between type of scenario, and revealed only the hypothesized main effect of situation, $F(1, 396) = 465.34$, $p < 0.001$. This reflects discounting of trait aggressiveness regardless of the scenario (see Figure 2, first panel). In contrast, the ratings of selfishness, motive to revenge and motive to gain money, showed a significant interaction, $Fs(1, 396) = 100.37$—1214.99, $ps < 0.001$. As can be seen in Figure 2 (second panel), consistent with the empirical data, although simulated selfishness was discounted in the revenge condition, it was not discounted in the instrumental aggression condition. For motives of revenge and gain (Figure 2, bottom panels), the simulation generally showed the same pattern of differences between conditions that encouraged and discouraged aggression like in the replication study.

6. General Conclusion

The present simulation was able to resolve an apparent paradox between trait and moral inferences found in my replication of [7]. By extending earlier connectionist architectures of person perception [10, 13] with the motives people may infer about the actor's behavior, discounting was consistently applied for trait ratings given various situational forces, but was absent for moral (i.e., selfishness) ratings given some types of situational forces. In other words, the traits ratings did not reveal the fundamental attribution bias, whereas the moral ratings did. A crucial feature that made it possible to simulate the differences between trait and moral inferences was that the selfishness and motive measures included either the motive to revenge or the motive to gain. This resulted in an asymmetric discounting pattern in the reactivity and instrumentality scenarios. In contrast, the trait measure did not involve these motives (although traits were affected by these motives during the learning history). Consequently, this resulted in a more symmetric discounting pattern across scenarios.

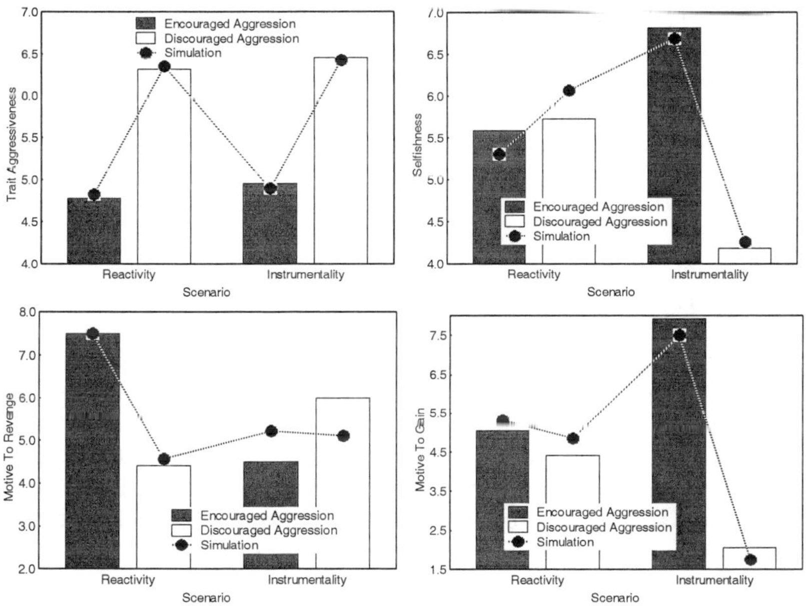

Figure 2. Perception of an aggressive actor as a function of scenario and situation. Observed and simulated values from the replication study.

References

1. Gilbert, D. T. (1989). Thinking lightly about others: Automatic components of the social inference process. In J. S. Uleman & J. A. Bargh (Eds.) *Unintended thought*. New York, NY: Guilford.
2. Kashima, Y, & Kerekes, A. R. Z. (1994). A distributed memory model of averaging phenomena in person impression formation. *Journal of Experimental Social Psychology, 30*, 407-455.
3. Kashima, Y., Woolcock, J., & Kashima, E. S. (2000). Group impression as dynamic configurations: The tensor product model of group impression formation and change. *Psychological Review, 107*, 914-942.
4. Kelley, H. H. (1971). Attribution in social interaction. In E. E. Jones, D. E. Kanouse, H. H. Kelley, R. E. Nisbett, S. Valins & B. Weiner (Eds.) *Attribution: Perceiving the causes of behavior* (pp. 1-26). Morristown, NJ: General Learning Press.
5. McClelland, J. L. & Rumelhart, D. E. (1985). Distributed memory and the representation of general and specific information. *Journal of Experimental Psychology, 114*, 159-188.
6. Read, S. J., & Montoya, J. A. (1999). An autoassociative model of causal reasoning and causal learning: Reply to Van Overwalle's critique of Read and Marcus-Newhall (1993). *JPSP, 76*, 728-742.
7. Reeder, G. D., Kumar, S., Hesson-McInnis, M. S. and Trafimow, D. (2002). Inferences about the morality of an aggressor: The role of perceived motive. *JPSP, 83*, 789-803.
8. Reeder, G. D., Vonk, R., Ronk, M. J., & Ham, J. (2003). *Dispositional attribution: Multiple inferences about motive-related traits*. Unpublished manuscript.
9. Shultz, T. & Lepper, M. (1996). Cognitive dissonance reduction as constraint satisfaction. *Psychological Review, 103*, 219-240.
10. Smith, E. R. & DeCoster, J. (1998). Knowledge acquisition, accessibility, and use in person perception and stereotyping: Simulation with a recurrent connectionist network. *JPSP, 74*, 21-35.
11. Van Overwalle, F. (1998). Causal explanation as constraint satisfaction: A critique and a feedforward connectionist alternative. *JPSP, 74*, 312-328.
12. Van Overwalle, F., & Jordens, K. (2002). An adaptive connectionist model of cognitive dissonance. *Personality and Social Psychology Review, 3*, 204-231.
13. Van Overwalle, F., & Labiouse, C. (2004). A recurrent connectionist model of person impression formation. *Personality and Social Psychology Review, 8*, 28-61.
14. Van Rooy, D., Van Overwalle, F., Vanhoomissen, T., Labiouse, C. & French, R. (2003). A recurrent connectionist model of group biases. *Psychological Review, 110*, 536-563.

APPROACHES TO EFFICIENT SIMULATION WITH SPIKING NEURAL NETWORKS

COLM G. CONNOLLY

Dept. of Computer Science,
University College Dublin,
Belfield, Dublin 4, Ireland.
E-mail: Colm.Connolly@ucd.ie

IOANA MARIAN

Autonomous Systems Lab, EPFL,
CH- Lausanne 1015, Switzerland.
E-mail: Ioana.Marian@epfl.ch

RONAN G. REILLY

Dept. of Computer Science,
National University of Ireland, Maynooth
Co. Kildare, Ireland.
E-mail: ronan@cs.may.ie

The distinct computational properties of spiking neural networks are increasingly the focus of research in computational neuroscience. When modelling these networks efficiency issues are critical. In this paper we present several algorithms for the event-driven simulation of spiking neural networks on single processor systems, which facilitate the simulation of large, highly active networks.

Computer simulations of the nervous system play an increasingly prominent role in understanding the way neurons process information. Spiking neural networks received special attention after experimental evidence accumulated to suggest that biological neurons use the timing of the spikes to encode and compute information (Thorpe, Fize, & Marlot, 1996).

Previous work on the efficient simulation of spiking neural networks indicated that high performance simulators for rate-coding neurons are not appropriate for spiking networks (Watts, 1994). Spiking networks, therefore require the design and implementation of their own dedicated environments.

In this paper we describe strategies for the time-efficient simulation of spiking neural networks on uni-processor machines. We begin by outlining a number of issues which must be addressed and then present a number of different approaches to economically managing inter-neuron communication.

1. Design considerations for efficient simulation

In this paper we will make use of a versatile spiking model known as the Spike Response Model (SRM) (Gerstner, 1999). Most SRM implementations use a continuous time model where the state of each neuron is updated at each time step (Fohlmeister, Gerstner, Ritz, & Van Hemmen, 1995). However, when a simulated neuron is reduced to a single variable, the effort devoted to updating a neuron's state can be equal to or exceed that of managing inter-neuron communication. It is this latter problem which must be adequately dealt with to economically simulate large spiking networks.

A number of other issues must also be addressed. Firstly, it is important to minimise the time spent integrating spikes. Secondly, the algorithm's efficiency depends on when each neuron is integrated. Finally, it is crucial, at the network level, to reduce the number of units that are integrated at once. In the following we discuss a number of ways to address these issues, namely, *how*, *when* and *how many units* to update.

1.1. *Optimisations at the neural level*

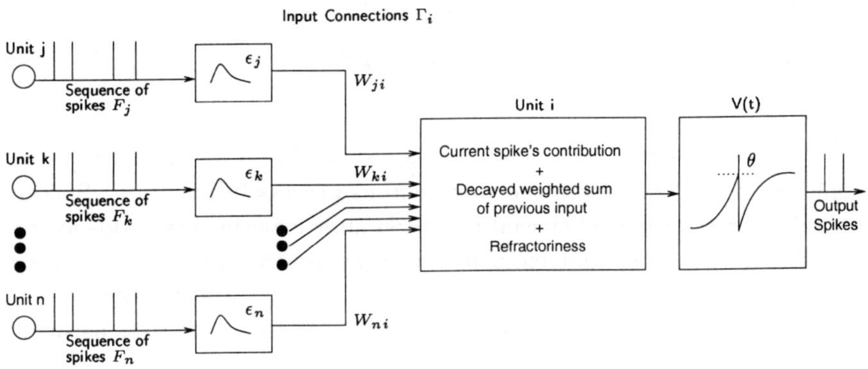

Figure 1. Representation of spike integration and generation in the spike response model. See in text for the meaning of the notation.

Sender-oriented connectivity. Since neurons in the brain are not fully connected (Braitenberg & Schuz, 1998), biologically plausible architectures should make use of neurons sparsely connected in a non-random fashion. Storage of these connections can be done using a list. A *sender-oriented* connectivity scheme is obtained if the target neurons are stored in the list. Otherwise, if the list stores the input neurons, it implements a *receiver-oriented* scheme. Previous simulation work suggested that a sender-oriented scheme is more appropriate for spike-processing networks (Marian, 2003).

Optimisation of the integration method. Another aspect to be considered is the integration of the membrane potential (see figure 1). The computational effort at this level derives primarily from the integration of synaptic inputs. The dynamics of neural activity merge stochastic and deterministic components, the first being associated with the random nature of the sequences of spikes and the second expressing the evolution of the neural state between two successive afferent spikes (Mattia & Del Giudice, 2000). Instead of computing at each time step, the weighted sum of the inputs, the integration method can exploit the deterministic evolution of the postsynaptic potential. The sum of past presynaptic spikes is stored and decayed every time the contribution of a new spike has to be added. Besides reducing the simulation time, this method also has better memory management properties (Claverol, Brown, & Chad, 2002; Marian, 2003).

1.2. *The Passage of Time*

The passage of time in neural simulators can be accomplished by two distinct methods. In the first of these, termed *continuous* or *time-driven*, the state of all the neurons in the network is updated at each time step. This is typical of simulators like GENESIS (Bower & Beeman, 1998) and NEURON (Hines & Carnevale, 1997).

In contrast, a *discrete* or *event-driven* system is one in which time is advanced in steps of variable size determined stochastically by the neurons in the network and where computation of the state of the neurons occurs only at the spike arrival time. This method best suits simplified spiking neuron models as they don't require the fine-grained advancement of time characteristic of continuous models. Spiking neural models, which incorporate suitable equations, usually allow all the computational effort to be focused on the time of arrival of spikes at a synapse. Consequently time can be advanced in variable sized steps. Since the time between spikes is usually random, the width of ΔT cannot be determined in advance. This means that when working with simplified spiking models the problem of simulating networks is reduced to one of managing the list of spikes generated by the neurons.

By adopting a simplified neuron model, choosing an efficient synapse storage mechanism, and adopting the discrete time model just described, the speed of the simulator will be predominantly determined by the queue management strategy. It is to queue management techniques, therefore, that we now turn.

2. Novel techniques for efficient simulation

It has been argued that event-driven strategies are efficient solely for the simulation of spiking networks with a low neural activity (Mattia & Del Giudice, 2000). However, our experience has not borne this out. We believe that judicious queue management can improve the efficiency of simulation even with active networks.

The most expensive operation in the event-driven simulation is insertion of events into the list sorted by delivery time. For the standard priority queue this operation is $O(\log_2(\text{length}(L)))$, where L is the length of the list. Thus, the longer the list, the longer the insertion time. This means that we require a method which can more efficiently manage spikes. In the following section, several strategies meant to improve the management of the event list are discussed.

2.1. *Reducing the number of events to be delivered*

The *multiple spike* method is a simple idea the main goal of which is to reduce the number of times that spikes are delivered to neurons. This is accomplished by aggregating all spikes to be delivered to the same target at the same time and delivering them together.

Implementation of this method can have an appreciable effect on simulation time. For example, if m spikes are to be delivered to a neuron the computational saving is $1/m$ since the neuron is updated only once instead of m times. The efficiency of this technique is greatly affected by the numeric precision of time. The larger it is the greater the probability of spikes aggregating and hence the fewer times neurons need to be updated.

Figure 2 shows results from four different simulations. This first of these is from a basic event–driven model. Here, each spike results in a neuron being updated. The next two are from models using the multiple spike

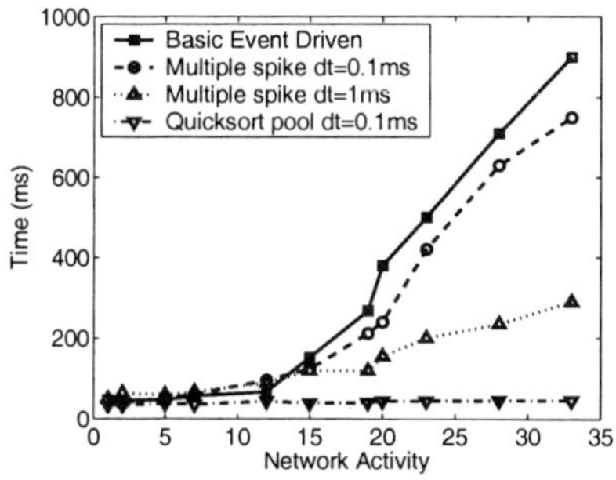

Figure 2. The computational effort, in CPU time, to integrate 1000 units vs. levels of network activity, when different event handling methods are applied. The network activity is measured as the average number of spikes in one ms divided by the total number of neurons, which in this case is 576. Computing a neuron involves: integrating its activity, generating output spikes with noisy delays, and inserting spikes in the event-list

concept. As can be seen, when the basic time step is larger, the amount of time taken to update the neurons is lower. This effect is more pronounced with greater network activity since the larger basic time step increases the probability of spikes aggregating and results in neurons being updated fewer times.

2.2. *Strategies for an efficient management of the events-list*

Under certain circumstances the multiple spike method may be inefficient. Thus there still remains the problem of efficiently managing spike lists. We considered four different solutions to this problem: layered delays-queue architecture, quick-sort pool, calendar queue and circular priority queue.

Layered delays-queue architecture. The approach taken by Mattia and Del Giudice (2000) is to split a single queue into several FIFO queues, each associated with a fixed axonal delay value. All synapses are organised in matrix-structured layers, each layer corresponding to one delay value. Adding a spike consists of appending it to the queue corresponding to the event's delay. Since all spikes in the same queue have the same delay the oldest will be at the top. Accordingly, the queues need no further sorting and insertion is done with O(1) complexity.

This method's major limitation is the number of synaptic delays. Although formal neural models simplify the biological neuron in several respects, most of the models account for noise effects on neural responses by introducing noisy thresholds, integration or delays (Gerstner, 1999). The transmission delays represent a new set of parameters that have no counterpart in traditional neural network models, and which can be used to read a temporal latency code (Thorpe & Gautrais, 1997) or to enhance the flexibility of spiking neurons used in learning algorithms (Gerstner, 1999). By discarding noisy synaptic transmission, this approach is suitable only for a limited set of applications.

Quick-sort pool. The solution to the efficient management of the events-list structure proposed by Marian (2003) eliminates the insertion overhead while preserving the essentials of neural behaviour (i.e., noise effects). Instead of performing the ordered insertion, a new spike-event is simply added to an unordered pool of spikes.

Time is divided into windows, T_{w}, and all spikes falling within a window are processed.

Since the spikes have to be processed in chronological order the simulator periodically stops and selects the spikes for the next window. This is done using a quick-sort algorithm. Most importantly, the sorting algorithm is run only over a small percentage of the elements in the pool, namely those whose time stamp falls within the next processing interval (see figure 3).

Shuffling of the pool is achieved by setting the first pivot of the quick-sort procedure to $t+T_{\mathrm{w}}$. The algorithm's efficiency depends on the choice of T_{w}. On the one hand, a small value, no larger than the minimum synaptic transmission delay, is necessary to ensure the chronological order of the spikes. On the other hand, a larger window reduces the number of times the queue is sorted and prevents the algorithm from failing to find any events within the corresponding interval.

This method was designed to deal with highly active networks and has been used to simulate a number of cognitive tasks (Marian, 2003). It's efficiency can be seen in figure 2 where, in combination with the multi-spike method, it scales very well with increasing network activity. The performance differences are mainly due to the management of the spike-events.

We compared the layered-delays and quick-sort methods with respect to the CPU time spent updating a given number of units. The comparison was facilitated by the fact that both algorithms have been used for the simulation of spike-driven learning processes of similar complexities. The layered delays-queue was used with excitatory and inhibitory integrate-and-fire neurons, spiking at 2Hz and 4Hz, respectively. The quick-sorting strategy was used with a self-organising feature map of excitatory and in-

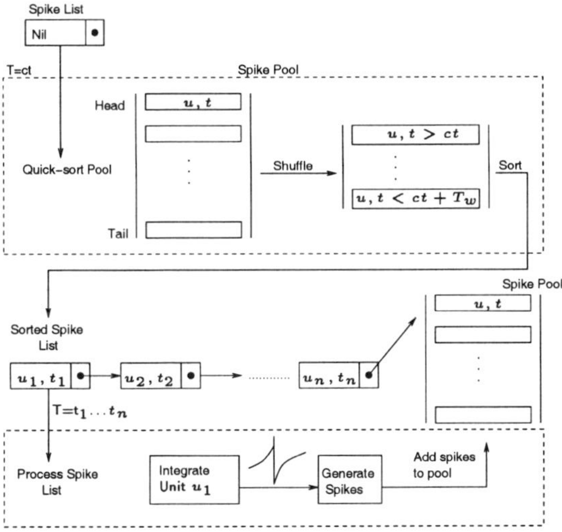

Figure 3. The event-driven simulation engine based on the quick-sorting of an unordered pool. Whenever the spike list becomes empty the quick sort algorithm is run upon a fraction of the elements in the spike pool, namely those with a time stamp between the current time t and $t + T_w$. Sorted spikes are stored in the list and are processed by the simulation engine. Any new events are simply added at the end of the pool.

hibitory neurons spiking at 100 Hz and with plastic excitatory and inhibitory synapses. The quick-sorting algorithm was tested for networks with the same size and connectivity rate (i.e. 10%) as reported by Mattia and Del Giudice (2000).

When comparing the performance of the algorithms, shown in figure 4(a), the huge disparity in the spike frequencies must be borne in mind. The quick sorting method manages to keep the simulation time approximately twice as long as the layered delays algorithm, even though it is processing spikes generated at 30 times the rate of those in the layered delays model. It does this and allows for an unlimited number of delays.

Calendar queues. Claverol et al. (2002) proposed a queue management strategy based on calender queues. This type of priority queue has fast O(1) insertion but O(n) removal. However, for queue occupancy rates greater than 3000, removal becomes O(1). The queue consists of a circular look up table (LUT) with each entry corresponding to a time step. All events with the same time stamp are stored together in the same entry in the LUT. Though similar to the multi-spike method described above, the events in the queue are not just spikes but messages of many sorts sent between the

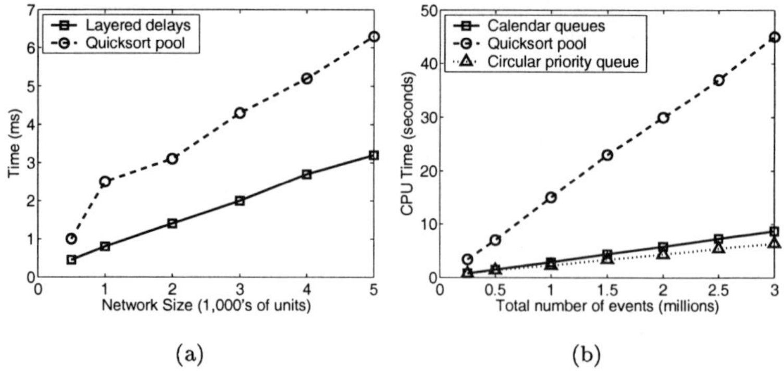

Figure 4. (a): Execution times per neuron vs. size of the network, computed for the simulation of the same number of firing units by two algorithms. (b): Simulation time versus total number of events or messages processed, for three different algorithms.

producers and consumers which make up the neural model. qq *Circular priority queue (CPQ).* This algorithm represents a combination of the multiple spike concept with the calendar queue technique. It consists of a two 2D matrices, with the rows corresponding to time steps and columns to neuron numbers. The CPQ incorporates the idea of the multiple spike by grouping all spikes pending delivery to a neuron on the same time step and delivering them together. The queue is circular, and wraps around every T time steps. A pointer to the head of the queue sets the current time step being processed and moving it from one row to the next advances time.

The function of the indirection matrix is to ensure that the order in which neurons are processed is different from one time step to the next. The column in the spike queue where a spike is stored is determined by using the spike's time and target neuron number as indexes into the indirection matrix. The row is determined by the time stamp of the spike. Once all spikes for a given time step have been processed the corresponding row in the indirection matrix is shuffled. Since the CPQ is based on calendar queues it has the same advantages and disadvantages, namely O(1) insertion and O(n) removal.

We compared the calendar queue, the quick-sort and the circular priority queue methods using the metric of time to process a given number of events. The results are shown in figure 4(b).

These demonstrate that the calendar type queues outperform the quick-

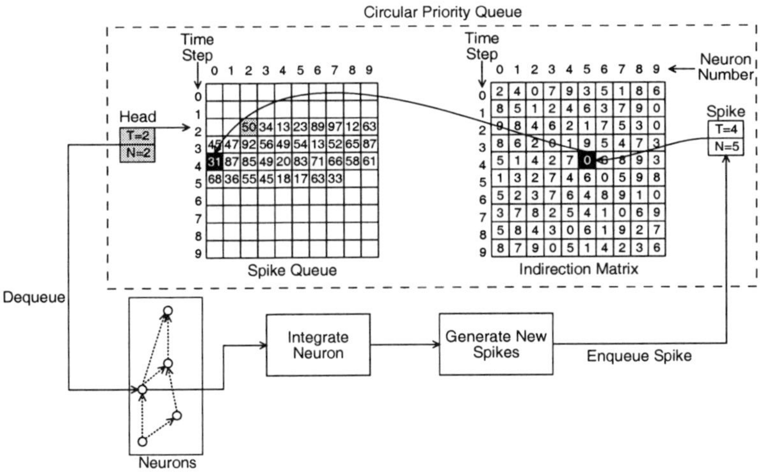

Figure 5. A circular priority queue incorporating the multiple spikes paradigm. Depicted is the queue for a network of 10 neurons. The queue length is 10, indicating the number of time steps, T, that the queue can store before it must wrap around. A value in the spike queue indicates the number of spikes pending delivery to a neuron. A new spike is stored in the spike queue in the column indicated by the indirection matrix and in the row corresponding to its delivery time.

sort method. The primary reason for this is that as the size of the queue grows, the quick-sort method must spend progressively more time sorting the spike pool. The methods based on calendar queues do not incur this overhead.

3. Summary

In this paper we have addressed the issues involved when simulating networks of simplified spiking neurons. Our primary concern is the efficient management of inter-neuron communication. We described several ways in which this may be accomplished. Exploiting determinism in the neural model, where possible, can focus the update of neurons solely on time of spike reception. This event-oriented approach can yield a significant improvement in simulation time and has become the de-facto standard for simplified spiking neural models.

References

Bower, J. M., & Beeman, D. (1998). *The book of GENESIS: Exploring realistic neuronal models with the GEneral NEural SImulation System* (2nd ed.). New York, NY: Springer Verlag.

Braitenberg, V., & Schuz, A. (1998). *Cortex: Statistics and geometry of neuronal connectivity.* Berlin: Springer-Verlag.

Claverol, E. T., Brown, A. D., & Chad, J. E. (2002). Discrete simulation of large aggregates of neurons. *Neurocomputing, 47*, 277–297.

Fohlmeister, C., Gerstner, W., Ritz, R., & Van Hemmen, J. (1995). Spontaneous excitations in the visual cortex: stripes, spirals, rings and collective bursts. *Neural Computation, 7*(5), 905–914.

Gerstner, W. (1999). Spiking neurons. In W. Maass & C. M. Bishop (Eds.), *Pulsed neural networks* (pp. 3–53). Cambridge, MA: The MIT Press.

Hines, M. L., & Carnevale, N. T. (1997). The neuron simulation environment. *Neural Computation, 9*(6), 1179–1209.

Marian, I. (2003). *A biologically inspired computational model of motor control development.* Unpublished master's thesis, University College Dublin, Department of Computer Science. (Available online at `http://cortex.cs.may.ie/theses/ioanaMarian/ioana_master.html`)

Mattia, M., & Del Giudice, P. (2000). Efficient event-driven simulations of large networks of spiking neurons and dynamical synapses. *Neural Computation, 12*(10), 2305–2329.

Thorpe, S., Fize, D., & Marlot, C. (1996). Speed of processing in the human visual system. *Nature, 381*, 520–522.

Thorpe, S. J., & Gautrais, J. (1997). Rapid visual processing using spike asynchrony. In *Advances in neural information processing systems* (Vol. 9, pp. 901–907). Cambridge, MA: MIT Press.

Watts, L. (1994). Event-driven simulation of networks of spiking neurons. In J. D. Cowan, G. Tesauro, & J. Alspector (Eds.), *Advances in neural information processing systems* (pp. 927–934). Morgan Kaufmann.

Language and Speech

READING, SUBLEXICAL UNITS AND SCRAMBLED WORDS: CAPTURING THE HUMAN DATA*

R.C.SHILLCOCK
University of Edinburgh
2 Buccleuch Place
Edinburgh, EH8 9LW
E-mail: rcs@inf.ed.ac.uk

P. MONAGHAN
University of York
York, YO10 5DD
E-mail: P.Monaghan@psych.york.ac.uk

We look at what role there might be for sublexical units in lexical representation in the modeling of isolated visual word recognition and in the reading of text. A variety of psycholinguistic paradigms have been used to investigate exactly how much information about letter order is required to recognize a word. We review some of the phenomena and some of the modeling solutions before suggesting an anatomically-based input representation that is capable of capturing important phenomena in reading in a parsimonious way.

1. Letter order information in reading words

1.1. *How much order information is necessary?*

> Aoccdrnig to a rscheearch at an Elingsh uinervtisy, it deosn't mttaer in waht oredr the ltteers in a wrod are, the olny iprmoetnt tihng is taht frist and lsat ltteer is at the rghit pclae. The rset can be a toatl mses and you can sitll raed it wouthit porbelm. Tihs is bcuseae we do not raed ervey lteter by itslef but the wrod as a wlohe. ceehlro.

The passage above was the subject of many thousands of emails in the autumn of 2003, and a corresponding level of media interest. The effect is self evident, although the explanation of the processing involved may go beyond the notion that "(we read) the word as a whole". In this chapter, we take this phenomenon as a starting point to discuss the general question of how the order of letters is

* This work is supported by The Wellcome Trust and by the Economic and Social Research Council (UK).

specified in reading, and to explore some potential modeling solutions to the problem of how the brain deals with such information.

The first point to note is that any discussion will depend on the identity of the language. The relevant behaviors, and any associated computational models, will vary substantially depending on the orthography involved. Most of our discussion will be restricted to English, but it is important to know if the behaviors generalize to non-alphabetic languages, such as Chinese, or to languages with radically different alphabetic orthographies, such as Hebrew with its right-to-left direction and its greater opacity resulting from the typical omission of explicit vowel information.

In English, to recognize a word like ORCHESTRA involves – at least – distinguishing that word from the 50,000 or so other words that a skilled reader of English may be expected to know. (Such a figure depends on the definition of "word", but the order of magnitude is correct.) Clearly, at least some information about the order of letters is necessary; if no such information were available, then it would be impossible to distinguish ORCHESTRA from CARTHORSE. However, it is by no means clear that the brain either requires or habitually uses all of the letter-order information that is available when a word is read.

Here we encounter the issue of the reader's purpose. At one end of the range, the reader may be proof- reading for typographical errors, in which case each word may be fixated, and some words may be fixated several times. At the other end of the scale, the reader may be scanning the text for the gist of the message it contains, in which case many words may not be directly fixated. In both cases, phonological information about the word(s) in the text may be mandatorily activated (see, e.g., Frost (1998) for further discussion) along with the meaning(s) present in the text. However, the reader's interaction with the words of the text is different, and we would expect, for instance, more spelling errors to be picked up in the first case, and we might expect the profile of the activation of the semantic representations associated with a word to differ according to whether the text was being read closely or not.

A second assumption is that the word is specified in terms of letters. There have been persistent attempts to show that the brain might store words somehow with reference to their overall shape. In fact, any such effects have only been shown in specific tasks such as proof-reading. The overall length of a word – the number of letters – is perhaps the most physically salient about a word but explorations of large databases of lexical decision times show that the length of a word is not a strong predictor of the time to judge that a stimulus is indeed a word. In normal visual word recognition the brain seems to process words in terms of letters, the most effective features for words in an alphabetic language (see Pelli, Farell & Moore, 2003).

For this discussion, we will assume that all that is required for word recognition is for the word to be differentiated from all of the rest of the words in the reader's mental lexicon. Previously, we have shown that a crude division of the word into just two "slots" is sufficient to uniquely specify a large proportion of words in a realistically large lexicon of English (Shillcock, Ellison & Monaghan, 2000). That is, CARTHORSE can be distinguished from the rest of the lexicon just by knowing A, R, C and T fall in the first half of the word, and R, H, O, S, and E fall in the second half; their precise order is not required. This fact should not be surprising for the longer words, but it is less clear that a very large percentage of shorter words can be so specified. In fact, only 4.7% of four-letter words from the CELEX English lexicon (Baayen, Pipenbrock & Gulikers, 1995) remain ambiguous when this basic left-right letter position information is known; these are words like TIME and ITEM.

In short, to distinguish one word from all of the others in the lexicon, we ideally need only know the identities of the letters in the right and in the left half of the word, and the problem is substantially solved.

However, almost 5% of four-letter words remaining ambiguous is not an acceptable level of success, so we must look for slightly more information about letter position than the minimal information we have so far considered. The ideal source of information comes "for free" from a longstanding observation that the brain seems to prioritize the processing of the outsides of words, i.e. the first and last letters (the "exterior letters") of English words.

This effect has been observed for short words for some time (see, e.g., Jordan, 1990). (There is some possibility that the effect will not transfer to longer words; if so, this might suggest that the behavior is a direct consequence of the informativeness of the exterior letters, rather than being due to the other potential explanations that have been advanced in the literature to account for the exterior letters effect.) Knowing the identity of the exterior letters of four-letter words necessarily disambiguates all of the four-letter confusions, like ITEM/TIME, and the only ambiguities that remain after specifying the identities of the letters in the left and right halves of the word, plus the exterior letters, are a tiny number of ambiguities like that between TRIAL and TRAIL.

This purely statistical analysis of the problem of word recognition makes the interesting prediction that the brain should be able to use its remarkable abilities in soft constraint satisfaction and content addressability to take a shortcut to identifying words based on this less specific information about letter position and order. The passage at the start of this chapter suggests that this claim is true (*pace* the presence of context).

A number of psycholinguistic experiments have been designed to explore this notion that a particular letter may contribute to the perception of a particular word even when it is outside its canonical position for that word. For instance, Humphrey, Evett and Quinlan (1990) have shown that BVK, but not TBVKU, primes the word BLACK.

In a related paradigm that stays within the perception of real words, Chambers (1979) and Andrews (1996) have shown that minimal pairs of words that differ only by two transposed letters, such as LOIN and LION, exert a mutual influence on each other in the lexicon. There are severe limits to the exploration of this effect in English as there are only a couple of dozen of such pairs of words in the language; in other languages, there may be more scope for further experiments to investigate the more precise parameters of this effect. However, even in English the effect is real: in a naming task, the more frequent word in the pair is responded to more slowly compared with a matched control word, and the less frequent word more quickly than its matched control word. If letters were processed strictly in their canonical position in the word, then LOIN should have no more effect on LION than LEAN, which also shares its first and last letters but has no strict match inside the word.

This broad class of experiments, and the transposed letters experiments in particular, represent a challenge to models of visual word recognition. They require that words be represented in such a way that letter position is not specified absolutely: a T in a particular position is required to have an effect outside that particular position. We consider now how computational modelers have responded to this challenge.

1.2. *Modelling solutions*

The computational modelling of visual word recognition, from Rumelhart and McClelland's Interactive-Activation Model onward, has made very substantial progress. In many ways, it has been the paradigm example of cognitive computational modeling: models have been implemented, predictions have been tested with human participants in the laboratory, and the models have been modified to produce further testable predictions, in a virtuous spiral. Furthermore, such models have provided an effective means of capturing and talking about the rich data from dyslexia. However, despite the progress made, computational models of visual word recognition can still be criticized from a number of perspectives.

First, there has been an almost exclusive concentration on monosyllabic words; even though the most influential models are able to employ all of the monosyllabic words of English, the paradigm "word" is still often the four-letter word of the IAM. Syllabic structure remains a problem in the mapping from English orthography to phonology.

Second, the starting point for modeling has typically been an abstract, computational construction of the problem, in which a word is seen as being composed of letters in specific positions. The simple four-letter-slot input of the IAM has remained an influential way of specifying the visual input.

Third, where modelers have departed from the simple letter-slot input, it has frequently been to replace it by one based on formal linguistic structure, dividing the monosyllabic input into onset, nucleus and coda: TRAM is represented as TR-A-M, in which there is no need to specify the ordering of the T and the R, as there is only one orthotactically legal order for them in an English onset. This move has been productive, guaranteeing good generalization by carving the problem at its joints, as it were. Nonetheless, there is a sense in which this pre-parsed input solves important aspects of the problem for the model before it starts.

Fourth, although the issue of distributed representations has dominated recent cognitive modeling, there has really been little further attempt to instill much resemblance to biological or anatomical reality in the models. (And the biological reality of the distributed representations used is itself rather abstract.)

Fifth, some modelers have specified the order of letters by specifying bigram units of various kinds (see, e.g. Mozer, 1987; Whitney, 2001). Although the details vary between models, hand-wiring such fine-grain parameters of a model runs the risk of producing a relatively brittle, language-specific architecture that fails to capitalize on emergent behaviors resulting from the precise lexicon used.

Sixth, in order to capture phenomena such as the ones we have dealt with in the first part of this chapter, some modelers have introduced new dimensions into their model, notably the idea of attention and attentional gradients (see, e.g., Whitney, 2001). Although such new structure in the models allows new representational freedoms, it is an expensive innovation. The nature and role of attention is controversial, and to allow attention to play a role in the construction of the representations of the words themselves, as opposed to applying once the words have become part of the perception, introduces more power and complexity into the models than may be necessary. If a more parsimonious model can capture the relevant data, then Occam's razor applies.

In our recent research, we have pursued a solution to the representation of the input to visual word recognition based on the observable anatomy of the visual pathways (see, e.g., Shillcock & Monaghan, 2001). The critical aspect of this anatomy is that there is a "vertical" division between the two hemispheres of the brain, and this division extends right out to the fovea. The human fovea is precisely vertically split, so that when a word is fixated, the part of the word to the left of fixation is initially projected to the right hemisphere (RH), and the part of the word to the right of fixation is initially projected to the left hemisphere (LH).

We can now refer back to our exploration of the statistics of visual word recognition, reviewed in the first part of this chapter. There we saw that simply knowing which letters are in which half of the word is a very substantial clue to the identity of the word. If that word has been fixated somewhere near the middle of the word, then we can see that the whole process of word recognition is starting with some very favorable information. There is no surer way of marking the location of particular letters in a word than by having them represented in different hemispheres of the brain.

Our first requirement of a model of visual word recognition is therefore that it should not throw away potentially useful information. In general, the brain's central expertise seems to be in detecting statistical regularity, and aspects of its architecture – in particular, the degree of recurrent connectivity and the division of the visual pathways into high and low spatial frequencies – seem to allow different levels of visual complexity to interact flexibly. It seems very likely that the brain does not throw away information about the hemifield of origin of particular letters. In our modeling, we might begin with the input CHA in the RH and IR in the LH (idealizing away from the splitting of individual letters), whereas other models begin simply with the input CHAIR. We are thus able to calculate letter position with respect to an anatomical given – the fixation point – as opposed to relying on word-internal *n*-grams, which might structure the word CHAIR into C_A, A_R, AI, CH, and so on.

There is a fundamental point here, concerning representation in the brain. The default view of how the visual input is treated is that a peripheral stimulus is subject to a series of transformations, becoming increasingly abstract and transcending its origins at the periphery: the physical representation of a word on the retina ends up as a lexical concept somewhere in the temporal cortex. However, such a unidirectional cascade of information, from the periphery to the centre, is in contrast to what we know about the pervasive recurrent connectivity in the brain. There is extensive recurrence back as far as the LGN, and the representation in V1 has to be intact for full, conscious awareness of the stimulus. Indeed, the representation in V1 is the largest, most authentic representation of the visual stimulus. And, crucially, it is divided.

In this view, there is no possibility of the brain losing sight of the original information concerning the hemifield of origin of the letter information. Figure 1 shows how these claims regarding input representation can be implemented.

	c	l	a	m
		c	l	a
			c	l
				c

m				
a	m			
l	a	m		
c	l	a	m	

Figure 1. The set of positions in which the word "clam" is fixated.

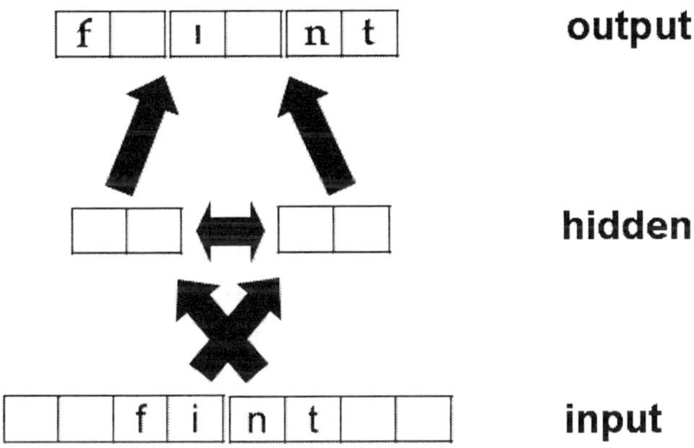

Figure 2. Schematic view of the split-fovea model.

In Figure 1, we see that the word can be fixated between any two letters (the middle three rows of the figure) or the fixation point can fall just before or just after the word (the first and last rows). This principled arrangement of the input is the one that we have used in the connectionist modeling of visual word recognition, in the split-fovea model, illustrated in Figure 2, in which each arrow represents a pattern of complete connectivity between layers of the model. The model is trained by recurrent backpropagation, over a series of time-slices, to associate the orthographic pattern at the input with the phonological pattern at the output.

Each row of Figure 1 represents the input to the model on a different occasion during training. It does *not* mean that a word has to be learned or perceived by fixating it at every possible fixation point successively. Rather, it simply reflects the fact that over a number of encounters with a particular word the reader will inevitably accrue a complete experience of that word, in which it straddles the fixation point – the vertical midline in Figure 1 – at every possible point. The brain must learn to deal with this variety of input representations of the same word.

At first sight this specification of the visual input may seem repetitive and elaborate, perhaps akin to McClelland and Elman's (1986) use of repeated templates of a single word to simulate the matching of a single representation against a temporal input. In the present case, however, the repetition is anatomically justified; the brain really does encounter words split in two in this manner. We have explored some of the behavior of a class of connectionist models trained on the divided input illustrated in Figure 1. The models themselves are divided, as in Figure 2, so that two sets of hidden units converge on a single output representation. (The output may be made to represent the orthographic or the semantic form of the orthographic input word, not just the phonological form, depending on the aspects of reading to be modeled.) The split models must generate internal representations that mediate the many-to-one mapping between, for instance, all of the input versions of CLAM shown in Figure 1 and a single phonological output for that word. The input representation we have described permits several things to happen.

First, it forces the model to develop adequate internal representations for every sublexical sequence of letters continuous with one or other end of the word. For monosyllabic words, such sequences will include all of the onsets, rimes and codas. For polysyllabic words, syllables and morphemes will be included in such sequences. Thus, the input representation potentially provides us with all of the advantages of the pre-parsed inputs seen in many current models of visual word recognition, but from a more naturalistic beginning.

Second, our explorations of such models show that the lexical identities of the type developed by connectionist models are preserved, even though the model is trained on multiple versions of each word in the shift-invariant mapping we have described. The models are able to learn the mapping between orthography and phonology to human levels of performance with words and nonwords. The lexical identity of irregular words is preserved, in that the model can learn their pronunciation in each possible position. In fact, similar mappings between orthography and context-vector semantic representations can also be learned (see Monaghan, Shillcock & McDonald, *in press*), where the mapping between orthography and semantics is very largely arbitrary. Success in learning such mappings is an even stronger demonstration that the model's exaggerated form of superpositional storage does not dissolve words simply into the representation of particular letters in particular positions.

Third, the shift-invariant mapping allows the model to concentrate on the information identified in our conventional statistical analysis of the problem of visual word recognition. We have already shown (Shillcock & Monaghan, 2001) that such split connectionist models robustly reproduce the exterior letters effect when the mapping is a simple one from orthography to orthography. In Figure 1, we can see the reason for this behavior: there are individual presentations of each word in which the first letter or the last letter is stranded in one or other hemifield and thus may command the resources of half of the entire model. This processing priority generalizes to the exterior letters in other positions. Furthermore, the model is able to take advantage of the lesser status of ordering information in the middle of words to generate intermediate representations that emphasize letter identity over letter order in those parts of the word, thereby capturing the human data that we considered in the first part of the chapter.

In conclusion, basing the model's architecture on observable anatomical constraints allows us to capture human reading behavior that itself reflects the demonstrable statistics associated with the problem of the recognition of isolated English words.

References

1. Andrews, S. (1996). Lexical retrieval and selection processes: Effects of transposed-letter confusability. *Journal of Memory and Language, 35*, 775–800.
2. Baayen, R.H., Pipenbrock, R. & Gulikers, L. (1995). The CELEX Lexical Database (CD-ROM). Linguistic Data Consortium, University of Pennsylvania, Philadelphia, PA.
3. Chambers, S.M. (1979). Letter and order information in lexical access. *Journ. Verb. Learn. Behaviour, 18*, 225–241.
4. Frost, R. (1998). Towards a strong phonological theory of visual word recognition: True issues and false trails. *Psychological. Bulletin, 123*, 71–99.
5. Jordan, T.R. (1990). Presenting words without interior letters: Superiority over single letters and influence of postmark boundaries. *Journal of Experimental Psychology: Human Perception & Performance, 16*, 893–909.
6. Monaghan, P., Shillcock, R. & McDonald, S. (*in press*). Hemispheric asymmetries in the split-fovea model of semantic processing. *Brain & Language.*

7. Mozer, M.C. (1987). Early parallel processing in reading: A connectionist approach. In M. Coltheart (Ed.), *Attention and Performance XII: The psychology of reading* (pp. 83–104). London: Erlbaum.
8. Pelli, D.G., Farell, B. & Moore, D.C. (2003). The remarkable inefficiency of word recognition. *Nature, 423*, 752–756.
9. Shillcock, R., Ellison, T.M. & Monaghan, P. (2000). Eye-fixation behaviour, lexical storage and visual word recognition in a split processing model. *Psychological Review, 107*, 824–851.
10. Shillcock, R.C. & Monaghan, P. (2001). The computational exploration of visual word recognition in a split model. *Neural Computation, 13*, 1171–1198.
11. McClelland, J. L. & Elman J. L. (1986). Interactive processes in speech perception: the TRACE model. In D. E. Rumelhart & J. L. McClelland Eds. *Parallel Distributed Processing,* Vol. 2, 58-121, Cambridge, Mass: MIT Press.
12. Whitney, C. (2001). How the brain encodes the order of letters in a printed word: The SERIOL Model and selective literature review. *Psychonomic Bulletin & Review, 8*, 221–243.

HOW THE CONSTRAINTS ON ENGLISH COMPOUND PRODUCTION MIGHT BE LEARNT FROM THE LINGUISTIC INPUT: EVIDENCE FROM SIMPLE RECURRENT NETWORKS

J.A. HAYES[1], V.A. MURPHY[1], N.DAVEY[2] AND P.M. SMITH[1]
Departments of Psychology [1]
and Computer Science [2],
University of Hertfordshire,
College Lane, Hatfield,
AL10 9AB, United Kingdom.
E-mail: J.hayes@herts.ac.uk

Native English speakers include irregular plurals in English noun-noun compounds (e.g. *mice chaser*) more frequently than regular plurals (e.g. **rats chaser*) (Gordon, 1985). This dissociation in inflectional morphology has been argued to stem from an internal and innate morphological constraint as it is thought that the input to which English speaking children are exposed is insufficient to signal that regular plurals are prohibited in compounds but irregulars might be allowed (Marcus, Brinkmann, Clahsen, Wiese & Pinker, 1995). In addition, this dissociation in English compounds has been invoked to support the idea that regular and irregular morphology are mediated by separate cognitive systems (Pinker, 1999). The evidence of the simple recurrent networks (SRNs) presented here is used to support an alternative view that the constraint on English compounds can be derived from the general frequencies and patterns in which the two types of plural (regular and irregular) in conjunction with the possessive morpheme occur in the input.

1. Introduction

1.1. *Compounding in English*

Psycholinguistic research has shown that English compound words with irregular plural nouns in first position (e.g. *mice-eater*) are produced far more frequently than compound words with regular plural nouns in first position (e.g. **rats-eater*), (Gordon, 1985).

1.2. *Explanation according to the Dual Mechanism model*

The dual mechanism model (Pinker, 1999), proposes that irregular nouns and their plurals are stored as memorised pairs of words in the mental lexicon (e.g. *mouse-mice*) but that regular plurals are produced by the addition of the [–s] morpheme to the regular stem at a post lexical stage (e.g. *rat* + *s* = *rats*). Compounds are created in the lexicon. Thus as irregular plurals are stored in the lexicon they are available to be included within compound words. However, as only the singular stems of regular nouns are stored in the lexicon the plural form is never available to be included within compound words (Marcus et al, 1995).

1.3. *A Single Route Associative Memory Based Explanation of Compounding*

An explanation of the constraints on English compounding based on the frequency of co-occurrence of items in the linguistic input has not been considered to date. It is argued here that children may not include the high type frequency regular [-s] plural morpheme in the middle of words such as compounds because they will always have heard it at the end of words. Thus, to include the regular plural morpheme [-s] in the middle of words contravenes a very clear pattern discernable from the input. Furthermore, frequency counts of a sample of the CHILDES (Child Language Data Exchange System) corpora (McWhinney & Snow, 1985) have shown that the plural [-s] morpheme is rarely followed by a second noun (Hayes, Murphy, Davey, Smith and Peters, 2002). Importantly, a different pattern is found with the possessive [-'s] morpheme since it is nearly always followed by a second noun. Therefore, it might be that a noun rarely follows the regular plural [-s] morpheme (i.e. patterns such as "**rat[s] chaser*" do not occur) because the pattern "*noun – morpheme[-s]-noun*" is reserved for marking possession (such as *rat's tail*). Interestingly in other languages that do not have this competition between the plural and possessive morpheme such as Dutch (Schreuder, Neijt, van der Weide & Baayen, 1998) and French (Murphy, 2000), there are examples of regular plurals being produced within compounds. Irregular plurals may, however, appear in English compounds as they are not formed by the addition of the plural [-s] morpheme. Thus, irregulars do not compete with the possessive structure and as such may be followed by a second noun in a compound even if in the case of mice or geese, for example, they end in an [-s] sound. The role of the plural [-s] morpheme as a predictor of word finality and the competitive relationship between the plural and possessive [-s] morpheme is examined here using a series of connectionist models.

2. Neural net modeling

2.1. *Experiment 1*

The Simple recurrent network (SRN) in Experiment 1, tested any role that [s] might play in indicating word finality in a stream of concatenated letters. The network was required to predict the next letter it expected to occur given the letters it had seen previously. At the beginning of a word the error was high but as more letters were presented to the network the error decreased until it was at its lowest at the end of the word. It was hypothesised that on a "next letter" prediction task of this kind, a neural network would learn that after the input [-s] there was a high probability that the next input would be a word ending marker. An SRN was used rather than a co-occurrence frequency analysis because it meant that the size of the training set utilised to make each individual prediction was decided by the network and not by the investigator. The network was trained on a concatenated stream of child directed speech taken from CHILDES (MacWhinney & Snow, 1985). Each letter of the alphabet plus a word ending marker was encoded using one of 27 random 5 bit vectors.

Results: The network's ability to learn that [–s] is a good predictor of word finality was tested using 19 unseen words that ended in [-s] and 19 unseen words that ended in other letters. The network was more accurate (i.e. the error was lower) at predicting a word ending marker after [-s] than after all other letters combined, $t\ (18) = -2.08$, $p = 0.05$.

2.2. *Experiment 2*

The objective of this SRN was to learn whether highly consistent patterns in the input (i.e. that a plural noun is rarely followed by another noun while a possessive noun is always followed by a second noun) can drive learning about how to manipulate plurals within noun-noun compounds. The network was required to predict the next word to occur given the words it had seen previously. Nouns, verbs, determiners and adjectives were used to make up legal English sentences. Similarly to the performance of Elman's (1990) model, this network was expected to be able to make a first order distinction between the function of the various syntactic types (learning that the words could be classified as nouns and verbs, determiners and adjectives). Furthermore, although the possessive and the plural [-s] were encoded in exactly the same manner in the input, it was predicted that the network would learn a second order distinction (a distinction that could only be learnt once the first order distinction had been learnt) that only "verbs" could appear after some [-s] morphemes and only "nouns" could appear after other [-s] morphemes. The network was trained on one group of nouns that were represented as having the properties of singulars, possessives and plurals (e.g. *hen*, *hens*, *hen's*). A second set was only represented as singulars and plurals (*coat*, *coats*), a third group was only represented as singulars and possessives (*wig*, *wig's*) and a fourth group was represented as singulars only (*bar*). Possessives and regular plural nouns were differentiated from singular nouns because they were encoded as ending in [-s]. A set of deverbal nouns were encoded by using the localist code for the verb (e.g., *drive)* and then representing the fact that they ended in the derivational affix [-er] (e.g., *driv(e) + e r = driver).* Nouns represented as singulars were followed by verbs ending in [-s] (to represent the third person singular). Nouns represented as plurals were followed by verbs that did not end in [-s]. Possessives were followed by deverbal nouns (i.e. verbs plus the the derivational morpheme [-er]). Apart from the fact that the singular, plural and possessive nouns were followed by items from different syntactic categories (as would be the case in the input the language learner is exposed to), the sentences that included the different types of noun were of similar length and construction. Thus the differences between plural and possessive nouns were only represented to the network by the fact that they were followed by items from different syntactic categories. It was predicted that items presented in the context of singulars, plurals and possessives would cluster together in the hidden layer representations.

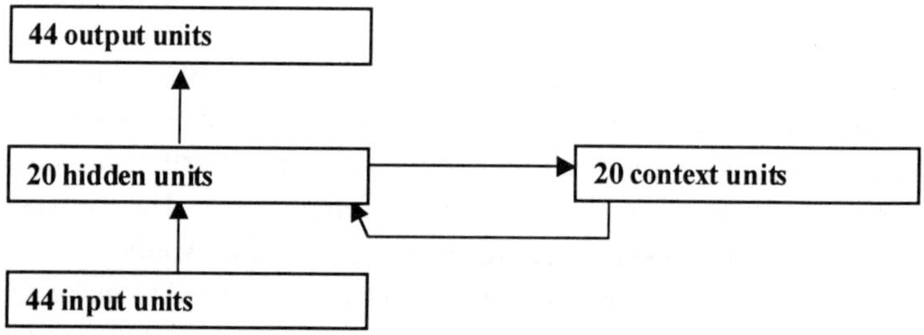

Figure 1. Architecture used for neural network 2

The network was trained on a concatenated stream of 2000 legitimate English sentences constructed from a lexicon of 37 words. Nouns with the properties of singulars, plurals and possessives were included with equal type frequencies in the input. A sentence-ending marker was attached to each sentence and the sentences (including the sentence-ending marker) were concatenated to form a stream of 14,600 words. Each word (including the sentence-ending marker) was encoded using a 38-bit localist coding scheme. The presence or absence of [-s] at the end of a word was also explicitly coded using 2 additional input units because the focus of this model was to investigate whether items ending in [-s] (phonetically identical items) could be differentiated depending on the co-occurrence patterns in which they occurred in the input. Whether the object of the sentence each word was included in was plural or singular was also encoded using 2 additional input units. Thus, for example, when the object of the sentence was singular noun *cat* the code would be *localist code for cat + [-s] off + plural off,* when the object of the sentence was plural noun *cats* it would be *localist code for cat + [-s] on + plural on* and when the object of the sentence was possessive noun *cat's* it would be *localist code for cat + [-s] on + plural off.* The presence or absence of [-er] at the end of a word was also explicitly coded using 2 additional input units.

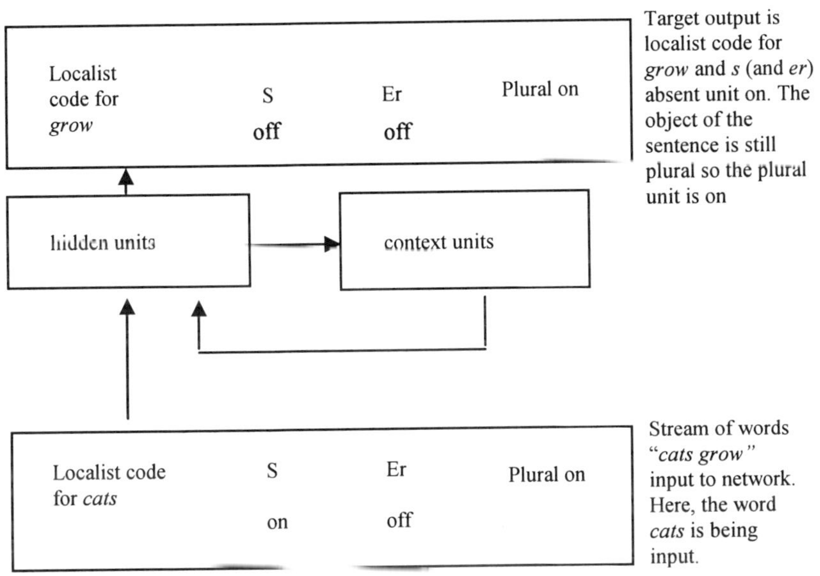

Figure 2. Diagram to represent word prediction task

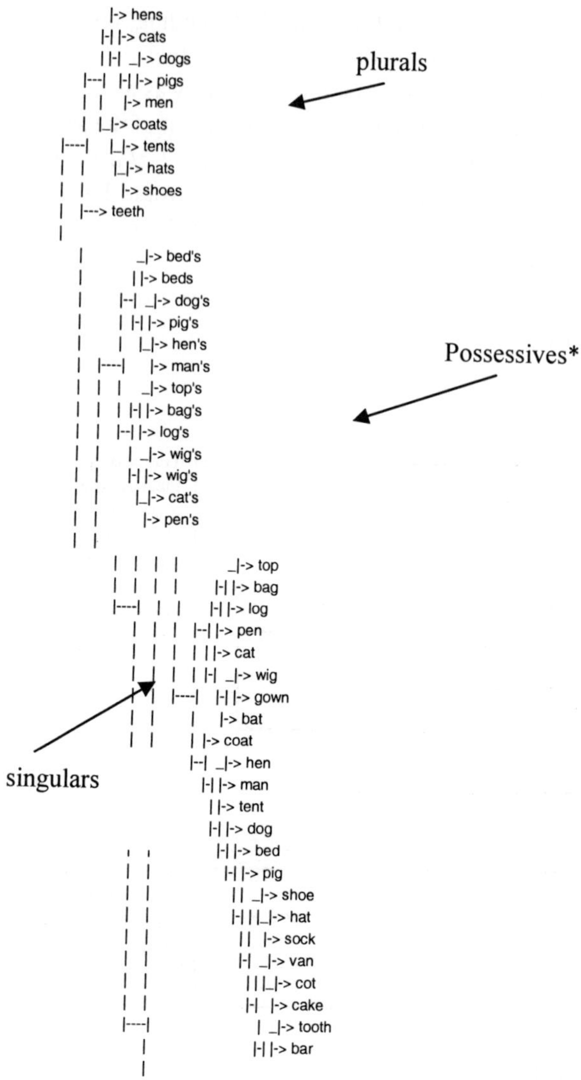

Figure 3. Cluster Analysis of hidden unit representations for model 2.
*The apostrophe is shown to distinguish possessives from regular plurals in this diagram but it was not represented in the training set.

Results: Figure 3 shows a hidden unit cluster analysis of the nouns used in the training set. It is evident that the network was very successful at clustering together nouns that performed similar functions in the training set. Words that appeared in the training set as both plurals and possessives (i.e. *hens/hen's cats/cat's, dogs/dog's, pigs/pig's, men/man's)* appear in both the cluster of possessive and the cluster of plural nouns indicating that something about their function in different contexts had been learnt by the network.

2.3. *Experiment 3*

This network was trained on a large training set of real child directed speech in which the frequencies with which the various types of morphology occurred were not manipulated in any way. The syntactic type of each word was used as the input to the network. The network was required to predict the syntactic type of the next input it expected to receive given the syntax of the previous inputs. The model in experiment 2 was capable of learning about grammatical type from a training set in which each word token was encoded using a localist coding scheme. In this earlier model items in the training set were not explicitly coded as being representatives of a particular syntactic type (e.g. as being nouns or verbs). Instead, learning about the distinct linguistic functions that the different syntactic types perform emerged during training. However, a disadvantage of this model was that it was only possible to use a small lexicon of words because of the complexity of the learning task. The model reported here was trained on a much larger training set than our previous model. This simulation sought to reproduce the behaviour of an older child, with a much larger vocabulary, who has knowledge, though perhaps not at a metalinguistic level, of the different functions that are performed by the different syntactic types.

The frequency in which regular and irregular plurals and possessives were included in the training set was determined by the frequency in which they appeared in the child directed speech that was used as the input to the model. The performance of the network was investigated using a syntactic type prediction task in which one of three syntactic types was input (a possessive, a regular plural or an irregular plural) and the network predicted which syntactic type it expected to see next in the input stream. The difference (error) between this predicted output and the output for noun, verb, other and word ending was calculated. It was predicted that the error would be high for all items after possessives except nouns. Conversely it was predicted that there would be a high error on predicting a noun after a plural of either kind.

Results: The error on producing the target output was recorded after the network was presented with the test sequences. Many runs of the simulation were carried out but each produced almost identical results.

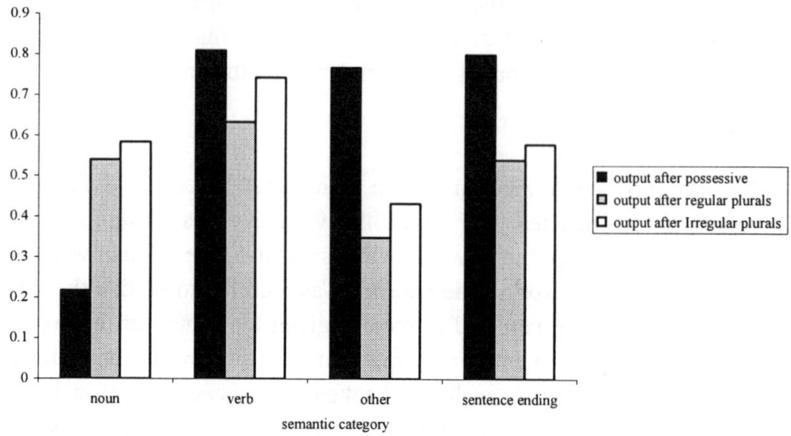

Figure 4. Error on producing nouns, verbs, other items and word endings after possessives, regular plurals and irregular plurals

Figure 4 illustrates that at a descriptive level the error on producing a singular noun after a possessive was about half as high as the error on producing a singular noun after a plural of either type. The network also learnt that the syntactic categories that make up other items and sentence-ending markers can follow plurals but not possessives[a].

[a] The network, produced a high rate of error when the target output after a plural noun was a verb, despite the fact that in the input verbs followed regular plurals (25% of the time that regular plurals occurred) and irregular plurals (11% of the time that irregular plurals occurred). However, the training set contained very few verbs
(6.24 % of the training set). Given that verbs were so underrepresented in the input it was unlikely that they would be predicted as the next item in a next word prediction task to any great extent.

3. Discussion

From experiment 1, it is evident that a neural net model trained on child directed speech was able to learn that [-s] is associated with word finality. This overwhelming pattern of [-s] at the end of words might influence language learners to omit [-s] from the middle of words. Experiment 2, showed that the model was able to learn that [-s] followed by one set of words was different from [-s] followed by a different set of words even thought the [-s] was encoded in exactly the same way in the input. The same might be true for the language learner. Both the possessive [-s] and the plural [-s] sound the same phonetically but the patterns in which the two different types of morpheme appear in the input may be enough to drive learning that one type of morpheme is used in some circumstances but not in others. In Model 3, syntactic category was explicitly encoded in the input and real child directed speech was used as input and thus the different syntactic categories were represented in the actual frequency that they occurred in real child directed speech. Under these more realistic input conditions there was still a suggestion that the network was able to recognise that the *noun –morpheme [-s]* pattern occurred in different patterns when it was plural to when it was singular. Specifically, the network showed some indication of being able to discern that nouns follow possessives but not plurals of either type and also of being able to detect that "other items" and word ending markers follow plurals of either type but not possessives. These three models taken together would seem to provide evidence for an associative account of compounding. In this associative account, the language learner notices that the [-s] morpheme tends to occur at the end rather than in the middle of a word. Furthermore, simply by exposure to the [-s] morpheme (i.e. without the plural or the possessive [-s] morpheme being explicitly labelled as being different from each other), the language learner becomes aware that the same [-s] morpheme occurs in different patterns in the input. With the addition of the absolute minimum of semantics, namely the numerical context in which the phrase is uttered, the language learner seems able to differentiate between the plural and the possessive morpheme. The possessive morpheme may be followed by a second noun but the plural morpheme may not be followed by a second noun. Thus when faced with a noun-noun compound the language user may delete the plural morpheme from the end of the first noun not because regular items of morphology are represented in a particular manner in the brain but simply because this pattern is used to denote possession not plurality.

References

1. Elman, J. L. (1990). Finding structure in time. *Cognitive Science, 14,* 179-211.
2. Gordon, P. (1985). Level-ordering in lexical development. *Cognition, 21,* 73-93.
3. Hayes, J.A., Murphy, V.A., Davey, N., Smith, P.M., & Peters, L. (2002). The /s/ morpheme and the compounding phenomenon in English. In W. D. Gray & C.D. Schunn (Eds.) *Proceeding s of the 24th Annual Conference of the Cognitive Science Society*. Mahwah, NJ: Lawrence Erlbaum Associates.
4. MacWhinney, B., & Snow, C. E. (1985). The Child Language Data Exchange System. *Journal of Child Language, 12,* 271-296.
5. Marcus, G. F., Brinkmann, U., Clahsen, H., Weise, R., & Pinker, S. (1995). German inflection: The exception that proves the rule. *Cognitive Psychology, 29,* 189-256.
6. Murphy, V. A. (2000). Compounding and the representation of inflectional morphology. *Language Learning, 50*, 153-197.
7. Pinker, S. (1999). *Words and Rules.* London: Weidenfeld & Nicholson.
8. Schreuder, R., Neijt, A., van der Weide, F., & Baayen, R.H. (1998). Regular plurals in Dutch compounds: Linking graphemes or morphemes? *Language and Cognitive Processes, 13*, 551-573.

USING THE STRUCTURE FOUND IN TIME: BUILDING DISTRIBUTED REPRESENTATIONS OF WORD FORMS BY ACCUMULATION OF EXPECTATIONS

F. MOSCOSO DEL PRADO MARTÍN,* R. SCHREUDER AND R. H. BAAYEN
IWTS, University of Nijmegen &
Max Planck Institute for Psycholinguistics
P.O. Box 310, NL–6500 AH Nijmegen, The Netherlands
E-mail: fermin.moscoso-del-prado-martin@mrc-cbu.cam.ac.uk

In this study we introduce the Accumulation of Expectations technique to build vectorial representations of the orthographic and phonetic forms of all the words in a language for use in connectionist models. We demonstrate how this technique can be used to build realistic orthographic representations for all Dutch and English words from the CELEX database, and realistic phonetic representations for all Dutch words in CELEX.

1. Introduction

The representation of the orthographic or phonetic forms of words to be used as input and output has long been a problem for connectionist models of language processing. The inherently sequential nature of human language is a factor that needs to be taken into account by paradigms to represent word forms. Some authors (e.g., Gaskell & Marslen-Wilson, 1997; Harm & Seidenberg, 1999; Plaut & Booth, 2000; Plunkett & Juola, 1999, Shillcock, Ellison, & Monaghan, 2000) have made use of templates that predefine slot-based templates for the sequence of letters or phonemes of which words consist. Such a technique limits the maximum length of a word to the number of slots representing letters, phonemes or morphological constituents that are predefined in the representation. Additionally, templates assume a predefined 'possible word' structure that requires pre-processing of the words in order to fit them into the templates. This introduces the additional problem of alignment by which, depending on the

*Current address: MRC–Cognition and Brain Sciences Unit, 15 Chaucer Rd., CB2 2EF, United Kingdom

goal of the model, the words need to be aligned to their beginnings, endings or word centers. This approach has been criticized for implicitly assuming symbolic processing. Other authors have tried different schemas, such as variants of the 'wickelgraph' (cf., Wickelgren, 1969) used in many models (e.g., Mozer, 1987; Rumelhart & McClelland, 1986; Seidenberg & McClelland, 1989), which introduce units corresponding to sequences of letters, phonemes, or phonetic features. These approaches are not only unrealistic, but also they are not capable of unambiguously representing all words in a language (Prince & Pinker, 1988). In our studies, we require a paradigm to represent unambiguously the formal properties of all words in a language. This paradigm needs to yield word representations that encode in a realistic way the form similarities between the words, and needs to be suitable to be used as input or output of backpropagation networks. Additionally, the similarity spaces must be created automatically, without manual preclassiffication of the different segments, and need to be language specific, to capture the language-specific way in which two words may be similar. Finally, a continuous measure of word similarity between the representations of pairs of words is required. For these reasons, we developed the Accumulation of Expectations paradigm. This paradigm produces representations of word forms that fulfil our requirements.

2. General Description of the Technique

Our approach to representing the orthographic and phonetic forms of words draws on the language-specific regularities that can be learned by a Simple Recurrent Network (SRN; Elman, 1990; Jordan, 1986; Norris, 1990). Elman (1990; 1993) showed that an SRN trained on predicting the next element (phoneme, letter, or word) on a linguistic sequence develops sensitivity to the possible sequences in a language. Both for the phonetic and the orthographic representations, we constructed SRN's similar to the ones described by Elman, and we trained them on predicting the next letter or phoneme in the sequence using the whole set of phonetic or orthographic forms in English or Dutch.

Once a network had learned about the regularities of phoneme or letter sequences in a particular language, we used the representations that were developed in its hidden layer to create the vectors corresponding to individual words. Ideally, the activation of the units in the hidden layer after the sequential presentation of all of a word's letters or phonemes should contain detailed information about all the previous letters or phonemes.

However, this is not the case. As the network is trained on predicting letter or phoneme sequences from a particular language, the limitation of representational space in its hidden layer drives the networks to represent only that information about the past items in the sequence which is potentially useful for predicting the upcoming elements. This has the beneficial effect of giving rise to language-specific generalizations, but comes at the cost of actually 'forgetting' many details about the items that were encountered in the past when these details are not particularly informative about the elements that are to appear in the future. For instance, consider the morphologically complex English word *disclosure*. During presentation of this word to a network that has been trained in letter prediction on English words, the activation values at the network's hidden layer change in three stages. After presentation of the letters 'd' and 'i' at the beginning of the word, the network assigns a high probability to 's' being the next letter to appear in the input. This is because *dis* is a common prefix in English. However, after recognition of this prefix, i.e., immediately after presentation of 's' to the network, the amount of uncertainty increases again because there are many stems that could appear after *dis*, thus making it impossible to predict the following letter to come. This uncertainty decreases again after presentation of the 'c', the 'l' and the 'o', at which point the network assigns a high probability to the 's' in concordance with the many English words that can have *close* as their stem. Note at this point that for the prediction of the 's' in *close* the network would not need much information about *dis* having appeared before it. Therefore, because of the limitation of representational space, most of the information about the prefix *dis* would have disappeared at the point when the network encounters the 'u'. A similar process takes place here, once it is known that the stem was *clos*, the network would assign high probabilities to the initial letters of the possible suffixes that *close*, or a verbal stem of the same type, can take, and does not need to store detailed information about the preceding string (*disclos*).

In order to avoid this problem, instead of taking the activation of the hidden layer after presentation of the last letter or phoneme of the word as a word's representation, we represent words as the average of the activation values in the hidden layer after the presentation of each of the word's letters or phonemes. This approach ensures that information about each particular letter or phoneme is considered in the final representation, while at the same time giving different, context-sensitive representations to all the elements in the sequence.

At this point we need to address a technical problem. The activation of the hidden layer always contains at least some residual information about its activation values at the previous time steps. Consequently, plain accumulation of the activation values in the hidden layer would result in the information about letters or phonemes that appear early in the sequence receiving more weight in the accumulated representation than the letters or phonemes that appear later in the sequence. This has the undesirable effect of creating representations that give more importance to word beginnings than word centers, and more importance to word centers than word endings.

The solution to this technical problem is straightforward: We need a weighted averaging process such that the relative importance of the activation values increases with time. In other words, we modulate the contribution of the hidden layer at each time step to the accumulated representation by giving a higher weight to elements that appear late in the sequence than to elements that appear early in the sequence.

3. Technical Specification of the Network used for Orthographic Representation

3.1. *Network Architecture*

The network that was used for orthographic representations consisted of twenty-six input and output units each corresponding to one letter in the alphabet. The output layer contained an additional binary unit representing the end of a word. The network's hidden and context layers consisted of forty units each.

3.2. *Network Training*

The networks were trained on letter prediction for all words in the CELEX database, that is, 297,690 words in the case of the Dutch network, and 154,063 words in the case of the English network. Letters were presented sequentially in the input by setting to one the value of the corresponding unit, while setting all other units to zero. At each time step, the network was trained to predict the next letter in the sequence (or the end of word) by activating the corresponding output unit. Errors were calculated using the divergence between the actual output of the network (after normalization) and the desired output (the next letter in the sequence). We trained the networks by the simple recurrent backpropagation through time algorithm with bounded momentum descent (Rohde, 1999),[a] with a learning rate of 0.04, a momentum of 0.9, and a weight decay of $1 \cdot 10^{-6}$.

Training proceeded by presentation of 10^6 words to the network. Words were chosen randomly, each having a probability of being presented proportional to the logarithm of its CELEX frequency. The activation values of the units of the context layer were reset to 0.5 after presentation of each word.

We trained two different networks, one on letter prediction for all Dutch words in the CELEX database, and the other on letter prediction for all English words in CELEX.

4. Technical Specification of the Network used for Phonetic Representation

4.1. *Network Architecture*

For the phonetic representations we built an SRN with fifteen input and output units each corresponding to one of the phonetic features described by Moscoso del Prado, Ernestus, and Baayen (in press).[b] The output layer contained an additional binary unit representing the end of a word. The network's hidden and context layers consisted of forty units each.

4.2. *Network Training*

We trained the networks on phoneme prediction for all 297,690 Dutch words in the CELEX database. Phonemes were presented sequentially in the input by setting to one the values of the units corresponding to their positive features, and to zero all remaining units. The training regime and parameters with which this network was trained were identical to those used in the networks for orthographic representations, with the only difference that we used the cross-entropy between the actual output of the network and the desired output features (the next phoneme in the sequence) as the error measure, instead of the divergence that we used in the orthographic

[a]The bounded momentum descent algorithm is a standard momentum descent where the pre-momentum weight step vector is bounded to a maximum modulus of 1.0. After the momentum is added, the resulting weight change vector is bound to a maximum modulus $1/(1 - \text{momentum})$. This permits using higher initial learning rates than using traditional momentum descent, resulting in less adjustments in the learning rate during training.

[b]These 15 phonetic features are a refined version of the phonological features described by Booij (1995).

network.

5. Building the Orthographic and Phonetic Representations

After training the networks, the representation for a word was formed by presenting the word letter by letter (or phoneme by phoneme) at the network's input, and accumulating the activation values in the network's hidden layer at each time step. For words shorter than three elements (characters or phonemes), the activation values in all time steps received equal weights. If the word had $l > 3$ characters, the activation values of the hidden layer at time step i was weighted by:

$$w_i = 1 + \frac{\left(\frac{l-3}{l}\right)^2 \cdot i^3}{l^4 + l^3}. \tag{1}$$

Before applying the weights, the values of the weights for all time-steps corresponding to a word were normalized to sum up to 1.0. The weighting schema was determined empirically, by testing the effect of different weighting schemas on clustering large number of words by prefix, suffix, and stem. The elements that appear at the end of the word receive a higher weighting than the rest. This increase in the weights given to the late elements is more marked in long words than in short words, and non-existent in words of three or less segments.

6. Evaluation of the Representations

In order to use these representations in computational models of morphological processing, we need to ascertain whether the information contained in the vectors is sufficiently fine-grained to distinguish between words that share an affix or stem. For each of the representations that we built, English and Dutch orthographic vectors, and Dutch phonetic vectors, we examined whether the representations of words sharing a prefix, a suffix, or a word stem are systematically similar. Figures 1 to 3 compare the orthographic and phonetic representations of English and Dutch words sharing selected prefixes, suffixes, and word stems, using Principal Components Analysis (PCA). Each figure plots two sets of words, with each of the sets being composed of words that share an affix or a stem, in the plane defined by the two first principal components obtained from a PCA on their vectors. Figures 1 and 2 show comparisons for orthographic representations of Dutch and English respectively. Figure 3 shows a clustering arising from phonetic representations of Dutch words. Note that Figure 3 contrasts the Dutch prefixes *on-* and *ont-* in their phonetic representations. These two prefixes are extremely similar, with the first one being a substring of the second. Therefore there is a great deal of overlap in their representations. This is not surprising given that there are many Dutch words (e.g., *ontegenzeggelijk* – "indisputably") that start with the prefix *on-* followed by a form starting with a 't', thus being exactly the same as if the word had had *ont-* as a prefix. Even in these very similar cases, the representations of words having these two suffixes are systematically different, as can be observed in the figure by the much coarser distribution occupied by the words starting with the prefix *ont-*. In fact, plain examination of the values of the first principal component by means of unpaired, two-tailed, non-parametric Wilcoxon rank sum tests with continuity correction, reveals differences in the distributions of both kinds of words, both for the orthographic representations ($W = 49488, p < 0.0001$), and for the phonetic representations ($W = 71963, p < 0.0001$).

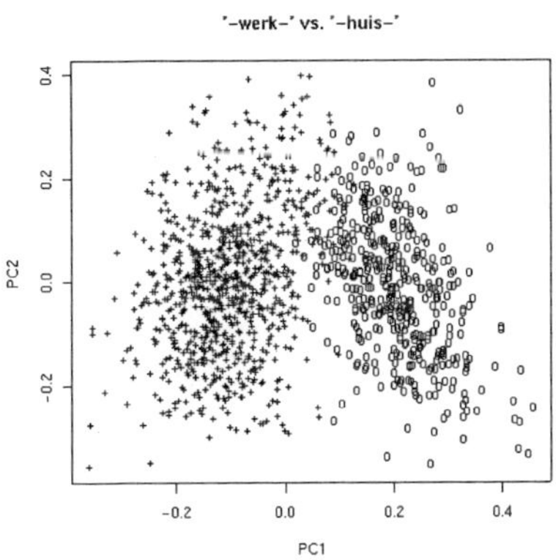

Figure 1. First two principal components contrasting the orthographic vectors of Dutch words that contain *-werk* (crosses) with those of Dutch words containing in *-huis-* (circles). The two first principal components explained 35% of the variance in this dataset.

7. Conclusions

In this study we have shown that the Accumulation of Expectations paradigm provides a useful method for representing orthographic and phonetic word forms in a way that can be used in connectionist models, over-

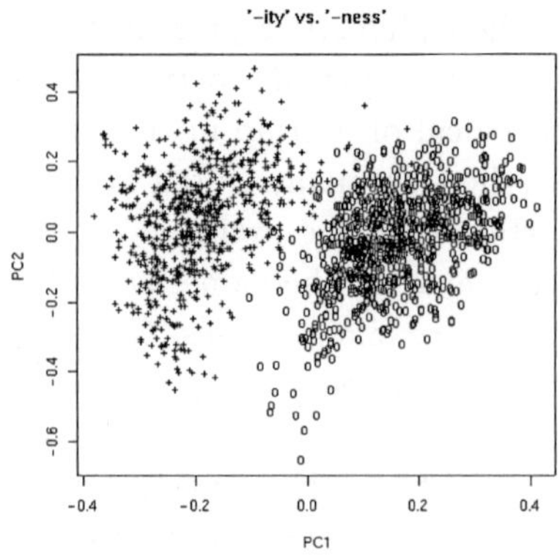

Figure 2. First two principal components contrasting the **orthographic** vectors of English words ending in the derivational suffix *-ity* (crosses) with those of English words ending in *-ness* (circles). The two first principal components explained 42% of the variance in this dataset.

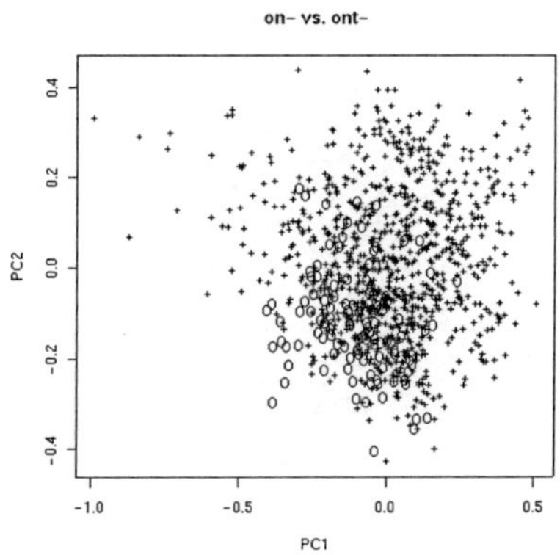

Figure 3. First two principal components contrasting the **phonetic** vectors of Dutch words starting with the Dutch prefix *on-* (crosses) with those of Dutch words starting with the Dutch prefix *ont-* (circles). The two first principal components explained 29% of the variance in this dataset.

coming the word length and preprocessing problems of previous template-based approaches. This paradigm extends the work of Elman (1990; 1993) on SRN's to cover all words in a language. Additionally, accumulating the activation values of the hidden layer allows us to create a single vector representation for every word in a language. This enables us to calculate the distance between two word forms using a traditional distance measure such as the cosine. We used SRN's to build word-form representations as they incrementally unfold, segment by segment. As a result we obtain a time-independent representation that can be used to reproduce a form in the output in a single time-step, while at the same time keeping track about the position dependent details of the segments. The Accumulation of Expectations technique provides representations that can be successfully used in connectionist models of language processing, as shown by the models described by Moscoso del Prado *et al.* (in press) and Moscoso del Prado (2003). Our technique also provides a continuous, language specific, measure of word similarity that is capable of predicting human responses in behavioural experiments, as shown by Dijkstra, Moscoso del Prado, Schulpen, Schreuder, and Baayen (submitted).

Acknowledgments

This work was supported by a PIONIER grant to the third author from the Dutch National Research Council (NWO).

References

Baayen, R. H., Piepenbrock, R. and Gulikers, L.: 1995, *The CELEX lexical database (CD-ROM)*, Linguistic Data Consortium, University of Pennsylvania, Philadelphia, PA.

Booij, G. E.: 1995, *The phonology of Dutch*, Clarendon Press, Oxford.

Dijkstra, T., Moscoso del Prado Martín, F., Schulpen, B., Schreuder, R. and Baayen, R.: 2003, Family size effects in bilinguals, *Manuscript submitted for publication, University of Nijmegen.*

Elman, J. L.: 1990, Finding structure in time, *Cognitive Science* **14**, 179–211.

Elman, J. L.: 1993, Learning and development in neural networks: The importance of starting small, *Cognition* **48**, 71–99.

Gaskell, M. G. and Marslen-Wilson, W.: 1997, Integrating form and meaning: A distributed model of speech perception, *Language and Cognitive Processes* **12**, 613–656.

Harm, M. W. and Seidenberg, M. S.: 1999, Phonology, reading acquisition, and dyslexia: Insights from connectionist models, *Psychological Review* **106**, 491–528.

Jordan, M. I.: 1986, Serial order: A parallel distributed approach, *Institute for Cognitive Science Report 8604*, University of California, San Diego.

Moscoso del Prado Martín, F.: 2003, *Paradigmatic Effects in Morphological Processing: Computational and cross-linguistic experimental studies*, MPI Series in Psycholinguistics, Max Planck Institute for Psycholinguistics, Nijmegen, The Netherlands.

Moscoso del Prado Martín, F., Ernestus, M. and Baayen, R. H.: in press, Do type and token effects reflect different mechanisms: Connectionist modelling of dutch past-tense formation and final devoicing, *Brain and Language*.

Mozer, M. C.: 1987, Early parallel processing in reading: A connectionist approach, *in* M. Coltheart (ed.), *Attention and Performance XII: The psychology of reading*, Erlbaum, London, pp. 83–104.

Norris, D. G.: 1990, A dynamic-net model of human speech recognition, *in* G. Altmann (ed.), *Cognitive Models of Speech Processing: Psycholinguistic and cognitive perspectives*, MIT Press, Cambridge, MA.

Plaut, D. C. and Booth, J. R.: 2000, Individual and developmental differences in semantic priming: Empirical and computational support for a single mechanism account of lexical processing, *Psychological Review* **107**, 786–823.

Plunkett, K. and Juola, P.: 1999, A connectionist model of English past tense and plural morphology, *Cognitive Science* **23**(4), 463–490.

Prince, A. and Pinker, S.: 1988, Wickelphone ambiguity, *Cognition* **30**, 189–190.

Rohde, D. L. T.: 1999, *A connectionist model of sentence comprehension and production*, PhD thesis, School of Computer Science, Carnegie Mellon University, Pittsburgh, PA.

Rumelhart, D. E. and McClelland, J. L.: 1986, On learning the past tenses of English verbs, *in* J. L. McClelland and D. E. Rumelhart (eds), *Parallel Distributed Processing. Explorations in the Microstructure of Cognition. Vol. 2: Psychological and Biological Models*, The MIT Press, Cambridge, Mass., pp. 216–271.

Seidenberg, M. S. and McClelland, J. L.: 1989, A distributed, developmental model of word recognition and naming, *Psychological Review* **96**, 523–568.

Shillcock, R., Ellison, T. M. and Monaghan, P.: 2000, Eye-fixation behaviour, lexical storage and visual word recognition in a split processing model, *Psychological Review* **107**, 824–851.

Wickelgren, W. A.: 1969, Context-sensitive coding, associative memory, and serial order in (speech) behavior, *Psychological Review* **76**, 1–15.

CONNECTIONIST MODELS OF SPEECH SEGMENTATION AND THE UTTERANCE BOUNDARY STRATEGY: A COMPARISON OF THE SOM, SRN AND N-GRAMS*

J. A. HAMMERTON
Alfa-Informatica,
University of Groningen,
Postbus 716, 9700 AS Groningen, The Netherlands
E-mail: james@let.rug.nl

Some connectionist models of speech segmentation have exploited the utterance boundary strategy, where the fact that utterance endings are also word endings is used to infer where word boundaries are. In this paper, it is demonstrated that using a simple N-gram based approach outperforms the neural networks for bigrams and especially for trigrams. Moreover the trigrams performance is better than that reported for all but the best 3 unsupervised models of speech segmentation in the literature. The implications of these findings both for connectionist models of segmentation and for the cognitive modelling of segmentation more generally are discussed.

1. Introduction

Spoken language has no equivalent of the spaces between words in written text. Thus the child learning language must learn how to split utterances into words without reliable pauses separating the words. The normal healthy child eventually becomes able to do this task with proficiency. Thus in modelling language learning, one must somehow explain how the child does this.

One proposal is that the child could exploit the fact that utterance endings are also word endings. Thus, when in the middle of an utterance they hear the same sounds they've heard at the end of an utterances before, they might posit a word boundary. This strategy, referred to here as the

*This work is supported by the Connectionist Natural Language Learning project of the High Performance Computing and Visualisation programme at the University of Groningen.

utterance boundary strategy (or UB strategy), can then provide a way of bootstrapping the problem of segmenting speech.

This paper takes two connectionist models of speech segmentation that exploit the UB strategy and compares their performance with an N-gram based algorithm that also exploits the UB strategy. The implications of this work are discussed.

2. Related work

In the sections below, references are made to "precision" and "recall", as measures of performance. Precision refers to the percentage of entities found by the learner that are correct. Recall refers to the percentage of correct entities that were found by the learner. The entities here are either words or word boundaries. So for example, if a precision of 50% is reported for finding words it means that of the words the learner found, 50% are correct. If a recall of 50% is reported, it means that 50% *of the correct words* were found.

Finally, the difference between finding a correct word and finding a correct word boundary is as follows. Finding a correct word involves finding 2 consecutive correct word boundaries. E.g. consider the sentence "The cat slept peacefully alone". If a segmenter produced "Thecat slept peacefully alone". It would have found 4 correct boundaries (including the end of the sentence), but only 3 correct words. The point is that finding a word correctly involves identifying both the start and the end of the word without there being any boundaries in between.

2.1. *Connectionist models*

Elman[5] provided an early connectionist model of segmentation, albeit of unsegmented text, using a simple recurrent network (SRN). This model exploits the fact that the letters at the start of a word will be less predictable than the letters within a word, but does not exploit the UB strategy. Elman trained a SRN with 5 inputs, 20 hidden units, 5 outputs and 20 context units for 10 iterations on an artificially generated set of sentences varying in length from 4 to 9, for a total stream of 1270 words and 4,963 letters. The network was trained to take a 5 bit representation of each letter (the representation of each letter generated randomly) as input and predict the next letter. Elman found that when he plotted the error curve of the trained network as it processed the sentences, the error at the start of a word was high and declined until the end of the word, arguing that this suggests that the input signal contains information about words. Brent[3] found that applying the Elman model to the Bernstein-Ratner corpus yielded a precision of roughly 30–40% and a recall of 40–45% for finding words.

Christiansen *et al* [4] present an SRN-based model, but here the UB strategy is employed. The network is trained to predict phonemes and UBs. The behaviour of the output unit used to indicate the prediction of a UB (referred to here as the UB unit) is used to predict word boundaries. Whenever the activation of the UB unit is above the average activation over the training set, a UB is predicted. The input to the network consists of phonemes, UBs (the network is not reset between utterances) and optionally indicators of stress. Christiansen took the Korman Corpus[9] corpus from the CHILDES project[10], and phonetically transcribed it using the MRC Psycholinguistic Database from the Oxford Test Archive. The training set thus consisted of 8181 utterances, with 927 test utterances from this corpus. The network achieved a precision of 37.31 and recall of 40.40 for finding words and a precision of 65.86 and 71.34 for finding word boundaries, using a SRN with 80 hidden/context units, phonemic information and UBs as input, a learning rate of 0.1 and a momentum of 0.9.

2.2. *Other models*

Brent[3] presents a model that segments utterances one at time, acquiring a lexicon in the process. Where the connectionist models above split their data into a training and testing set, this system is an incremental learner that learns as it goes. The model is called INCDROP and works qualitatively as follows. Segment each utterance in turn so as to optimise the following criteria:

(1) Minimise the sum of the lengths of all hypothesised novel words in the segmentation.
(2) Minimise the number of hypothesised words in the segmentation.
(3) Maximise the product of the relative frequencies of the words in the segmentation. The relative frequency of a word is the number of times the word has occurred so far as proportion of the total number of times all words have occurred so far.

Thus as INCDROP receives an utterance, it segments it according to the current lexicon and the above criterion, hypothesising new words where necessary. The INCDROP criteria can be derived from a probabilistic generative grammar encoding the prior knowledge that sentences are constructed by selecting words from some finite, but initially unknown lexicon and stringing them together. MBDP-1, an implementation of the INCDROP model, achieved a segmentation accuracy (in terms of correctly matched words) ranging from around 62% after processing 500 sentences to almost 80% after processing 9500 sentences from the Bernstein-Ratner corpus[2] from the CHILDES project[10]. The figures for recall ranged from 45% to just over 80%. For the corpus as a whole, Venkataraman[11] reports precision (accuracy) of 67 and recall of 69.39 for MBDP-1 for finding words. Note that INCDROP does not exploit the UB strategy and finds the optimal segmentation out of all possible segmentations given its current lexicon.

Batchelder[1] also presents a model which learns to segment speech and build a lexicon incrementally, called Bootlex. Where INCDROP splits utterances, Bootlex starts with the individual phonemes in an utterance and then clusters them. Bootlex operates as follows:

(1) Initialisation. Initially, the lexicon contains the set of phonemes, each with its own entry and a frequency of 1.
(2) The first utterance is parsed into word tokens of one symbol each, based on the initial lexicon.
(3) For each word token in the utterance just parsed, the corresponding word type in the lexicon has its frequency incremented by 1.
(4) Before the next utterance, the lexicon is augmented by adding to it new words consisting of consecutive pairs of words in the utterance just parsed. Each pair that is not in the lexicon is added to the lexicon with a frequency of 1.
(5) The second utterance is parsed into words, using the words in the lexicon, and a score for each possible parse is computed from its likelihood in light of the experience to date, using the frequency counts recorded in the lexicon. The word tokens from the highest scoring parse are used to update the frequency counts in the lexicon (step 3 above) and to make new entries (step 4 above).
(6) For all remaining utterances, step 5 is repeated, each time using the lexicon as just modified.

When Bootlex was applied to segmenting the Bernstein-Ratner corpus, it achieved a word precision of 67.2% and a word recall of 68.2%.

Venkataraman[11] provides a model similar to INCDROP but with a simpler probabilistic model underlying it. INCDROP's probabilistic model assumes the corpus is a single event in probability, where Venkataraman's model drops this assumption. Nevertheless Venkataraman achieves competitive performance with a precision of 67.7 and recall of 70.18 for finding words on the Bernstein-Ratner corpus.

3. Two connectionist models employing the UB strategy

3.1. *A model based on the Self-Organising Map (SOM)*

Hammerton[6] presents a model of speech segmentation based on the Self-Organising Map (SOM). It is assumed the reader is familiar with the SOM but Kohonen[8] provides a description. In order to process sequential information the standard SOM had to be modified. Instead of simply presenting each pattern in turn to the SOM, the inputs are updated as follows:

$$I(t) = I(t-1) \times D + P(t)$$

Where $I(t)$ is the input at time t, $P(t)$ is the pattern to be presented at time t and D is a decay constant, set here to 0.5. The inputs behave similarly to a leaky integrator and, given binary patterns, will preserve sequential information during processing. At the beginning of a new sequence, the inputs are reset. Hammerton used the SOM for speech segmentation as follows:

- The SOM was trained, for 1 iteration, on the training set.
- After training, the training set was presented one more time, and the units active at the ends of utterances noted. These units are now referred to as UB units.
- During testing, whenever one of the UB units is activated, a word boundary is predicted.

The idea is that the units active at the end of utterances will be sensitive to sequences of phonemes that form utterance endings and thus indicate word endings. The training set consisted of 4822 sentences from the Korman corpus of the CHILDES database and the testing set consisted 2610 sentences. Two representations for phonemes were used at the input layer; an orthogonal representation where one unit is one indicating which phoneme is being input and a representation where 2 units are on, and the rest off, with one unit indicating whether a consonant or vowel is presented and the other which phoneme it is. The model achieved a precision/recall of 55.63/87.94 for finding boundaries and 24.66/38.97 for finding words in the test set, when using a 40x40 map and a representation distinguishing consonants and vowels.

Here, this model will again be used, but will be compared with an SRN-based model (described in Section 3.2) similar to that of Christiansen *et al's.* Also the test data here consists of every 10th sentence extracted from the Korman corpus and phonetically transcribed using Celex. Where a sentence contained words not in Celex, it was discarded. The Korman corpus contained a total of 13350 sentences, of which 9644 were transcribed, and 8679 used as training data with the remainder (965) as test data.

3.2. *The SRN model*

Again it is assumed readers are familiar with the SRN, but Elman[5] provides a description. The SRN model here is trained using phonemic information at the input layer. At the output layer, the network predicts the next phoneme to appear or an utterance boundary as appropriate. The SRN is reset on each utterance. The same representations for phonemes were used at the input layer as for the SOM model, but the output layer always uses an orthogonal representation for the phonemes. The learning rate was 0.1 and there was no momentum.

Note that Christiansen's SRN model was not reset between utterances, had utterance boundaries presented as inputs and used momentum of 0.9. However the SRNs used here became saturated and didn't train if such a high momentum was used. Also, the SOM was trained to reset on each utterance, thus direct comparability required the SRN to do so too.

4. N-gram based segmentation using the UB strategy

The N-gram based algorithm works as follows. The training data is processed so that the final N phonemes of each utterance are stored in a simple lookup table, with duplicate N-grams removed. If an utterance has fewer than N phonemes, the whole utterance is memorised.

When processing the test data, each utterance is processed phoneme-by-phoneme by checking whether the N-gram formed by the current phoneme and the previous $N-1$ phonemes matches one of the N-grams memorised in the lookup table. If there is a match a word boundary is proposed after the current phoneme. This algorithm employs the UB strategy in a direct, simple manner, and thus forms a useful point of reference for other models that employ the UB strategy.

Table 1. Results for the SOM model, in terms of finding boundaries/words.

Net	Rep	Precision (b/w)	Recall (b/w)	Fscore (b/w)
32x25	orth	52.83/23.60	97.47/43.52	68.51/30.60
32x25	cons	56.43/26.76	95.57/45.32	**70.96/33.65**

5. Experiments

In all the experiments with the SOM, the initial learning rate was 0.5, reduced to zero at the end of training. The initial neighbourhood covered the entire map and was reduced to one unit at the end of training. Training involved a single presentation of the training data to the network. With the SRN, training also involved a single presentation of the training data, the learning rate was set to 0.1, and there was no momentum. All the figures reported here for the SOM and SRN are averaged results from 5 runs from different randomised weight initialisations.

Table 1 gives results for training various versions of the SOM model. The columns are as follows:

- "Net". This gives the size of the SOM used. E.g. 32x25 means there were 800 units organised on a 32 by 25 grid.
- "Rep". This indicates the representation used for the phonemes. "orth" means orthogonal, and "cons" indicates that consonants and vowels are distinguished.
- "Precision (b/w)". The percentage of predicted boundaries/words that are correct.
- "Recall (b/w)". The percentage of the correct boundaries/words that were predicted.
- "Fscore (b/w)". This is 2$\times$ Precision$\times$ Recall/(Precision+Recall).

Note that a word is regarded as being correctly predicted only when 2 consecutive boundaries are correctly predicted. The best performance came from using the representation distinguishing consonants and vowels.

Table 2 gives the results for various versions of the SRN model. Columns are as before, except "Net" indicates the number of hidden and context units. Where an "h" appears after the number it indicates that hysteresis was used in the context layer with a value of 0.5. Normally, the context layer simply holds a copy of the hidden layer from the previous time-step. With hysteresis, the context layer is updated as follows:

$$C(t) = C(t-1) \times h + H(t-1) \times (1-h)$$

Where $C(t)$ is the activation pattern of the context layer at time t, h is the

Table 2. Results for the SRN model in terms of finding boundaries/words.

Net	Rep	Precision (b/w)	Recall (b/w)	Fscore (b/w)
50	orth	66.27/32.29	72.21/35.18	69.11/33.68
80	orth	69.16/35.91	72.47/35.91	70.77/36.75
80C	orth	60.22/23.48	71.53/27.90	65.38/25.50
100	orth	69.27/36.06	71.49/37.22	70.35/36.63
200	orth	67.68/34.37	70.89/36.00	69.25/35.16
800	orth	57.05/24.82	70.02/30.47	62.86/27.35
100	cons	66.29/38.30	73.60/42.35	69.73/**40.30**
100h	orth	70.17/37.83	71.93/38.78	71.03/38.30
100h	cons	70.24/39.18	73.96/42.25	**72.05**/40.19

Table 3. Results for the N-gram model in terms of finding boundaries/words.

N	Precision (b/w)	Recall (b/w)	Fscore (b/w)
1	44.04/14.23	99.90/32.28	61.13/19.75
2	64.09/38.59	94.07/56.64	76.24/45.90
3	81.90/56.87	82.46/57.25	**82.18/57.06**
4	89.86/51.26	62.99/35.93	74.07/42.25
5	91.93/48.43	56.44/29.74	69.94/36.85

hysteresis value, and H(t) is the activation pattern of the hidden layer at time t. Note if $h = 0$ then this is the normal update of the context layer.

Thus instead of the context layer doing a straight copy of the hidden layer, the context layer and hidden layer pattern at each step are averaged together. This has the effect of improving memory capacity. As can be seen the best performance (figures in bold) comes from using hysteresis and/or the representation distinguishing consonants and vowels.

Finally, where a C appears after the number of hidden/context units it indicates that Christiansen's training regime was used, i.e. the context layer is not reset after each sequence, and utterance boundaries are presented explicitly at the input layer. As can be seen this diminished performance significantly. This suggests that retaining information in the context layer across sequences harms performance.

Table 3 gives the results for the Ngram models. Columns as above, except the first column gives the number of phonemes in each N-gram. The best performance comes from the tri-grams but both bi, tri and 4-grams outperform all the neural nets.

6. Discussion

Section 5 demonstrates that bi, tri and 4-grams outperform both connectionist models, and that the tri-grams' performance is only surpassed by that reported by INCDROP, Bootlex and Venkataraman's model. When applied in an incremental manner to the Bernstein-Ratner corpus for direct comparison with these models, the trigrams achieved a word precision/recall of 55.10/50.62, and a boundary precision/recall of 84.00/77.18. The boundary precision/recall here are comparable to Bootlex's 81/83.

The poor performance of the NNs compared to the N-grams is a challenge for connectionist approaches. If such a simple system, based on the same strategy the NNs are exploiting, can do so much better then why bother with the NNs? Some possible explanations for the networks poor performance are suggested here:

- Lack of resources. A lack of memory capacity or power in the networks may be the problem. However a range of sizes of network were tried and the best performing size used in both cases and using hysteresis with the SRN to boost memory only added a couple of points or so to the Fscore. But maybe applying a network with a capability for holding onto information for arbitrary time periods would help further, e.g. Long Short-Term Memory[7]. Also, finding the optimal decay rate for the inputs to the SOM may shed some light too. Finally, lack of training might be an issue, but note that Hammerton[6] found that, for the SOM, training for 400 iterations made only a few points difference on the Fscore.
- The networks are learning a different function from the N-grams. The SOM tries to find an optimal placement of the map units in input space to reflect its topology, whereas the SRN tries to predict the next phoneme or whether a UB is imminent. Thus neither attempt to memorise which sequences of phonemes form utterance boundaries. Note that if you trained a feed-forward network with a sliding window of inputs, you are unlikely to get the same performance as the Ngrams. This is because the activation of the UB unit is what determines whether a word boundary is predicted. If an Ngram appears frequently in the corpus but only infrequently as an utterance boundary, the UB unit may receive a below average activation despite the fact that the Ngram does sometimes appear as an utterance boundary. This may help explain why the SRN performs poorly compared to the Ngram strategy.

However the findings here raise issues for all the models of speech segmentation. Only the most sophisticated models, which acquire a lexicon and find the optimal segmentation of an utterance given that lexicon, are doing better than the N-grams.Yet the N-grams achieve performance not far below the sophisticated models. This raises the question of whether the models need to be so sophisticated.

Acknowledgments

Thanks go to John Nerbonne for discussions of and comments on this work, the participants at the 8th Neural Computing and Psychology Workshop who provided useful feedback, Morten Christiansen for answering questions about his work and Michael Brent for providing the means of generating the phonemic transcriptions of the Bernstein-Ratner corpus. Thanks also to an anonymous reviewer for some constructive comments.

References

1. E.O. Batchelder. Bootstrapping the lexicon: A computational model of infant speech segmentation. *Cognition*, 83:167–206, 2002.
2. N. Bernstein-Ratner. The phonology of parent child speech. In K. Nelson and A. van Kleeck, editors, *Children's Language*, volume 6. Lawrence Erlbaum Associates, Hillsdale, NJ, 1987.
3. M. Brent. An efficient, probabilistically sound algorithm for segmentation and word discovery. *Machine Learning*, 34:71–106, 1999.
4. M. H. Christiansen, J. Allen, and M. Seidenberg. Learning to segment speech using multiple cues: A connectionist model. *Language and Cognitive Processes*, 13:221–268, 1998.
5. J. L. Elman. Finding structure in time. *Cognitive Science*, 14:179–211, 1990.
6. James Hammerton. Learning to segment speech with Self-Organising Maps. In Tanja Gaustad, editor, *Computational Linguistics in the Netherlands 2002: Selected Papers from the Thirteenth CLIN Meeting*, pages 51–64. Rodopi, Amsterdam, 2003.
7. S. Hochreiter and J. Schmidhuber. Long short-term memory. *Neural Computation*, 9(8):1735–1780, 1997.
8. T. Kohonen. The self-organising map. *Proceedings of the IEEE*, 78(9):1464–1480, 1990.
9. M. Korman. Adaptive aspects of maternal vocalizations in differing contexts at ten weeks. *First Language*, 5:44–45, 1984.
10. B. MacWhinney. *The CHILDES project: Tools for analyzing talk.* Lawrence Earlbaum Associates, Mahwah, NJ, third edition, 2000.
11. A. Venkataraman. A statistical model of word discovery in transcribed speech. *Computational Linguistics*, 27:351–372, 2001.

Cognitive Architectures and Binding

DESIGNING AN OSCILLATORY MODEL OF BRAIN COGNITIVE FUNCTIONS

R.M. BORISYUK
Centre for Theoretical and Computational Neuroscience
University of Plymouth
Portland Sq, Plymouth PL4 8AA, UK
E-mail: rborisyuk@plymouth.ac.uk

Y.B. KAZANOVICH
Institute of Mathematical Problems in Biology, Russian Academy of Sciences
Pushchino, Moscow Region, 142290, Russia
E-mail: kazanovich@impb.psn.ru

We describe the principles of designing an oscillatory neural network for processing visual information and show that the functioning of the system can be based on two main principles: the synchronization of oscillators via phase-locking and resonant increase of the amplitudes of oscillators if they work in-phase with other oscillators. The processing of a printed word is presented as an example of computer simulation.

1. Introduction

A traditional approach to visual pattern recognition is based on the assumption that objects are presented one at a time, but in reality biological systems must be able to deal with visual scenes that contain several objects simultaneously. Experimental studies show that a number of cognitive functions are in action when analyzing complex scenes. Firstly, the whole information should be collected in the pools belonging to different objects and the background. This separate representation must be preserved in the further process of feature extraction and transformation, which ensures a proper conjunction (binding) of features at the level of object recognition and memorization. Secondly, attention is necessary to decrease the amount of information that is processed in detail and to improve the binding of features. Thirdly, pattern recognition and memorization should be combined with novelty detection to ignore familiar or unimportant objects.

Until now most papers on neural networks have been focused on modelling a particular cognitive function. Models of feature extraction and binding have been developed in [8, 22, 26]. Models of attention are represented by both traditional connectionist networks [7, 10, 21] and oscillatory networks [5, 12, 13, 16]. Models of memorization and novelty detection can be found in [4, 18, 25].

These models paved the way to combining in one system a set of cognitive functions covering the whole range of visual image processing.

Below we describe a large-scale model that includes the main stages of information processing in the visual pathway: (1) segregation of information from different objects according to their spatial connectivity; (2) consecutive selection of objects into the attention focus; (3) extraction of features and their transformation to the form invariant to object location and scale; (4) representation of objects in the working memory and novelty detection.

Presuming that the brain does not invent a special procedure for each cognitive function but adapts similar mechanisms for a particular type of processing, it has been a challenge to develop a model that would rely on a small set of general principles. As such principles we use oscillations, synchronization, and resonance. The choice of an oscillatory neural network for the development of the model is conditioned by the fact that animals and humans display a wide spectrum of rhythmic activity patterns in many areas and structures of the brain (see, e.g., [2, 15]) and that oscillatory principles provide efficient means for many types of information processing [3].

2. The Model

The model is designed as a hierarchy of interactive modules. Each module consists of oscillators with synchronizing or desynchronizing connections. An oscillator used as an element of the network is described by three variables: the oscillation phase, the oscillation amplitude, and the natural frequency of the oscillator [4, 13]. The values of these variables change in time according to prescribed rules of interaction. These rules include phase-locking, resonant increase of the amplitude of oscillations and adaptation of the natural frequency of an oscillator to the current frequency of oscillations.

The flow of information between the modules of the network is presented in Figure 1. The components of the network are called Object Selection Module (OSM), Local Feature Module (LFM), Invariant Feature Module (IFM), and Novelty Detection Module (NDM). The oscillators comprising these modules are denoted as OSO, LFO, IFO, and NDO, respectively. There is also an additional Central Oscillator (CO) with global feedforward and feedback connections to the OSM. The top-down interaction is reduced to the one from the NDM to the OSM (its role is explained in Section 2.1). More sophisticated top-down interaction is postponed for future versions of the model.

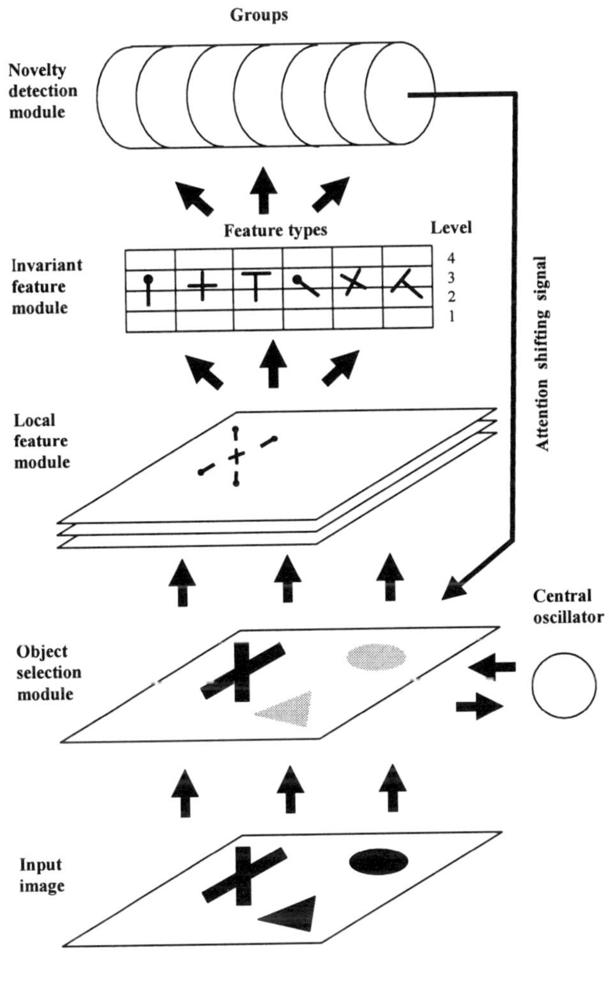

Figure 1. The architecture of the network. The input image contains three objects. In the OSM an object in the focus of attention is painted in black, other activated regions are painted in gray. In the LFM and IFM, there are five features of the object in the attention focus: four endpoints of different orientation and a crossing of two lines. In the LFM the features are attached to special locations where they are found, in the IFM the features are registered independently of their location in the image. The NDM is divided into the groups located along the horizontal axis.

In biological terms the model is interpreted in the following way. It is assumed that oscillators in OSM, LFM, and IFM represent cortical columns in the areas of the visual pathway. The OSM is located in the primary visual cortex (striate cortex), LFM can be attributed to different regions of the cortex (striate, extrastriate and higher) depending on the type and complexity of the features,

IFM represents feature detectors of the temporal area invariant to geometrical transformations (IT and higher associative areas). The CO plays the role of a central executive of the attention system [1, 6]. To simplify the model, we consider only the interaction between the CO and OSM and ignore the influence of the CO on other modules. The NDM is associated with the hippocampus [14]. The groups of NDOs represent hippocampal segments.

2.1. *Object Selection Module*

The OSM is responsible for grouping the information from the external input into separate clusters according to spatial connectivity of objects. The OSM has the same 2D grid-like structure as the visual field with one-to-one correspondence between the pixels of the image and the elements of the module. An OSO with coordinates *(x, y)* is activated by the input signal from the pixel *(x, y)*. The grouping of pixels into object representation is realized through synchronizing local connections between OSOs.

The OSM is also used to organize consecutive selection of objects into the focus of attention. This is achieved through phase-locking interaction of OSOs with the CO (see [13] for details). The interaction is organized in such a way that at any moment the CO works coherently with an assembly of OSOs that represents a single object. This is achieved through the interplay between synchronizing influence of OSOs on the CO and desynchronizing influence of the CO on the OSOs. The synchronizing connections from the OSM to the CO result in phase-locking of the CO by an assembly *A* of OSOs. In addition, the CO adapts its natural frequency to reach perfect tuning with the oscillators in *A*. On the other hand, the desynchronizing connections from the CO to the OSM prevent simultaneous synchronization of the CO with several unconnected assemblies of OSOs.

Due to the resonance with the CO, the amplitude of oscillations in *A* is made high while the activity of other OSOs is inhibited to a low level. Being in the resonant state is interpreted as the fact that *A* is included in the focus of attention.

The resonant state in *A* is interrupted by the top-down signal from the NDM that is generated when an object in the attention focus is detected as a new one and memorized in the working memory or when the familiarity of the object is detected. This signal blocks the assembly of OSOs so that it is unable to interact with the CO until the whole image is analyzed. This gives the CO an opportunity to change the attention focus by synchronizing its activity with another assembly of OSOs, etc. The order in which objects are included in the focus of attention is determined by their saliency. Objects of greater size and contrast have an advantage in being attended first.

2.2. *Local Feature Module*

The LFM is responsible for transforming the information about an object from representation by pixels to representation by local features. The oscillators in the LFM are arranged into a 3D structure with different types of feature detectors occupying different layers (planes). An LFO is active if a corresponding local feature is present in the object that is currently in the attention focus. The assembly A is used as a common source of synchronization for active LFOs. The filtering of signals outside the attention focus is determined by the amplitudes of the signals coming from the OSM. Since at each moment the features of only one object are included in the attention focus, this filtering allows the system to prevent erroneous conjunction of features of different objects at later stages of information processing.

2.3. *Invariant Feature Module*

The IFM is used for representing an attended object by a set of features that are invariant to object transformations such as translation, scale, etc. The module is arranged as a set of K columns of oscillators, where K is the number of types of feature detectors in the LFM. An IFO at the level i in the column k is active if a local feature of the kth type is present i times in the attended region of the input image. Thus at most one IFO can be active in a column at any moment. Such a coding automatically makes the activity in the IFM invariant to translation. Invariance to scaling can be achieved if the set of features extracted from an object is independent of the scale. We will give an example of such features in Section 3.

The assembly of synchronous LFOs plays the role of a common source of synchronization for all active oscillators in the IFM. In this way the synchronization that has appeared in the OSM is spread to higher modules generating a representation of an object in the form of coherent oscillations.

2.4. *Novelty Detection Module*

The NDM is responsible for memorization of objects in the working memory and making decisions about novelty of objects. The discrimination between novel and familiar objects is made in terms of duration of oscillatory activity in the NDM in response to external stimulation by a visual object. Following the experimental evidence on the theta activity in the hippocampus during orienting response [19, 23], the NDM generates a long (tonic) response when an object is new and a short (phasic(response when an object is familiar. The memorization of an object in the NDM is achieved under fixed connection strengths by a proper modification of internal parameters of NDOs. The details of this modification can be found in [4].

The NDM is an elongated structure divided into independent (disconnected) groups of oscillators located in the planes orthogonal to the long horizontal axis. There are all-to-all synchronizing connections between NDOs in each group. Connections from the IFM to the NDM are of all-to-all type with random delays. These delays mimic phase lags in transmission of the signals from the neocortex to the hippocampus.

A basic principle of NDM functioning is that an NDO reaches and keeps a high level of activity (resonant amplitude) if the signals that are supplied to this oscillator from the IFM arrive in-phase. For a given set of active oscillators in the IFM, due to random delays in connections, the resonant activity in the NDM takes place at only a small number of randomly chosen locations (groups), where an appropriate coincidence of input signal phases takes place. The activity in other parts of the NDM is low. Thus each object is represented in the NDM by a sparse assembly of oscillators that is specifically related to the object in the attention focus.

The activity in the NDM is organized so that under the influence of the coherent input from the IFM the number of resonant NDOs gradually increases until it reaches a certain threshold H, that is until the assembly of resonant oscillators in the NDM becomes sufficiently large. At that moment the NDM generates the top-down signal to the OSM that leads to the shift of the attention focus to another object. The important parameter is the period of time Δt from the moment when attention is focused on a given object and until the assembly of at least H resonant oscillators in the NDM is formed. By a proper modification of parameters of NDOs during memorization, it is possible to accelerate their capability to generate the resonant activity, therefore Δt can be made much smaller for familiar objects than for novel objects [4].

3. A Simulation Example

We illustrate the principles of network performance using a simple black and white image representing the characters of the word "HELLO". The image is exposed at the input for 35 time units and is processed sequentially object by object with memorization and novelty detection of all 5 objects. In this example the order in which objects are selected is conditioned by their size.

Since the object L occurs two times in the image, it should be detected as familiar at the second appearance in the attention focus. Other objects occur in the image only once, therefore they will be detected as new.

Fig. 2 shows the types of features used to represent the image HELLO in the LFM. For example, the object H is represented by six active LFOs - four endpoints (two top and two bottom) and two T-shape crossing (left and right). In the IFM this character is represented by four active IFOs – top endpoint (level 2), bottom endpoint (level 2), left T-shape crossing (level 1), and right T-shape crossing (level 1).

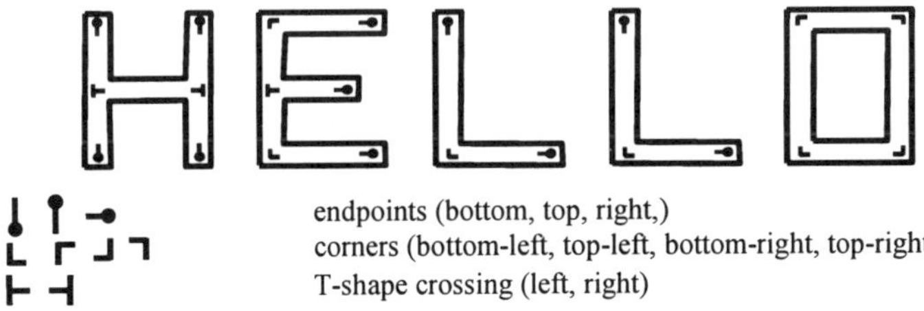

Figure 2. The features used for coding the shape of objects in the image.

Figure 3 shows the dynamics of the amplitudes in the OSM. By this figure one can see the periods when different objects are attended which is reflected in high amplitudes of oscillations of the OSOs corresponding to these objects while the amplitude of other OSOs is low. The period of time Δt when an object is attended varies between 4.1 and 5.0 for a new object. For a familiar object (the first appearance of L in HELLO) the value of Δt is 1.6 which is about 3 times shorter.

4. Discussion

We have demonstrated that general principles of information processing in oscillatory neural networks can be successfully applied to the solution of complex cognitive tasks that combine several interrelated cognitive components such as feature binding, attention, and novelty detection. The system architecture and functionality reflect (in a very simplified form) the main stages of visual information processing, starting from the primary visual cortex and finishing at the hippocampus. By computer simulations we have shown that the system is capable to fulfill consecutive selection of objects in the image and their novelty detection in terms of the duration (tonic or phasic) of the oscillatory response at the output module (the hippocampus).

The main new aspects of the model are the implementation of selective attention and novelty detection. Separately these cognitive functions have already been modeled in our previous works [4, 13], now we provide a framework where both models can be properly adjusted to each other. The consecutive selection of objects in the attention focus is crucial for functioning of the modules IFM and NDM. In the IFM it prevents the erroneous computation of invariant features by combining local features of different objects. In the NDM it prevents the memorization and novelty detection of chimerical combinations of several objects.

The principles of information processing used in our system have already appeared in other models. Our achievements are mostly related to their proper combination and adaptation to the task considered. As far as modeling the binding problem, we follow the already known ideas (see [3] for a review), reformulating them in terms of oscillators with explicitly defined phase. The peculiarity of our model is that we use the characteristics of individual pixels as primary features. The advantage of this approach is that it can be applied to any type of images and not only to contour objects.

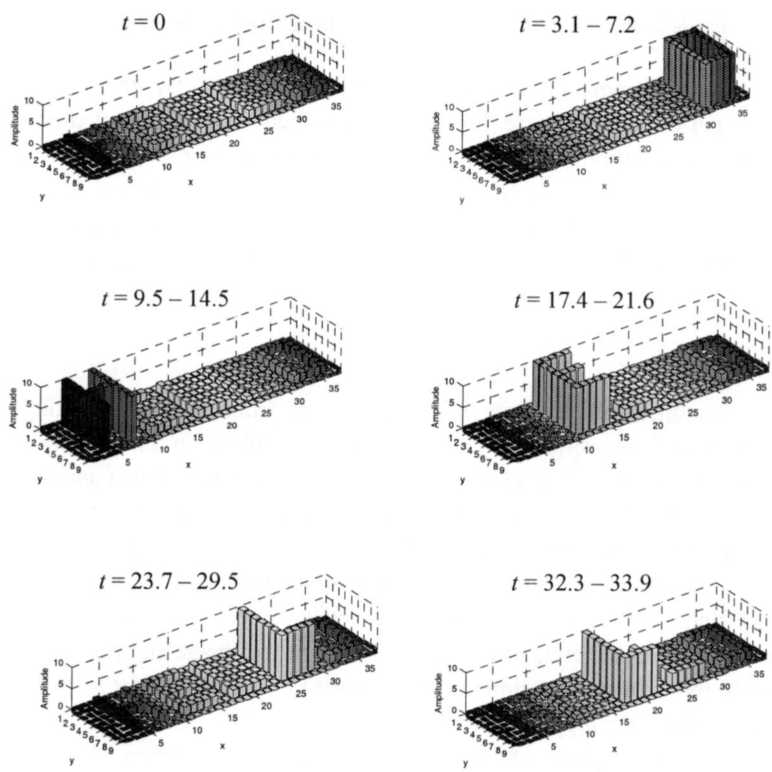

Figure 3. Dynamics of the amplitudes in the OSM. The numbers above the graphs show the period of time when attention is focused on a given character.

The idea of the resonant interaction is very attractive and finds support in experimental and modeling studies [9, 11]. Our approach differs from the one developed in these works because we explicitly postulate the type of dynamics of the oscillator amplitude depending on the synchronization with other oscillators.

The adaptation of natural frequencies of oscillators has been used before as a mechanism of learning and memorization [4, 17, 20]. Here it provides an efficient mechanism of implementing a winner-take-all procedure when different assemblies of oscillators compete for the synchronization with the central oscillator.

Oscillatory models of attention with the central element have been developed in the papers [24, 26], where the role of a central element in the network LEGION is played by a global inhibitory neuron, and [5], where the central element is represented by a population of integrate-and-fire neurons. The function of the central element in these works is similar to the one considered here, that is to synchronize some assemblies of oscillators and to desynchronize others.

5. Acknowledgments

This work was supported by the Russian Foundation of Basic Research (Grant 03-04-48482 and Grant for Scientific Schools 1872.2003.4).

References

1. Baddeley, A. (1996). Exploring the central executive. *Quarterly Journal of Experimental Psychology, 49A*, 5-28.
2. Basar, E. (1998). *Brain function and oscillations*. Springer, New York.
3. Borisyuk, G., Borisyuk, R., Kazanovich, Y., & Ivanitskii, G. (2003). Models of neural dynamics in brain information processing – the developments of 'the decade'. *Physics – Uspekhi, 45*, 1073-1095.
4. Borisyuk, R., Denham, M., Kazanovich, Y., Hoppensteadt, F., & Vinogradova, O. (2001). Oscillatory model of novelty detection. *Network, 12*, 1-20.
5. Corchs, S. & Deco, G. (2001). A neurodynamical model for selective visual attention using oscillators. *Neural Networks, 14*, 981-990.
6. Cowan, N. (1988). Evolving conceptions of memory storage, selective attention and their mutual constraints within the human information processing system. *Psychological Bulletin, 104*, 163-191.
7. Grossberg, S. & Raizada, R. (2000). Contrast sensitive perceptual grouping and object-based attention in the laminar circuits of primary visual cortex. *Visual Research, 40*, 1413-1432.

8. Hummel, J.E. (2001). Complementary solutions to the binding problem in vision. *Visual Cognition, 8*, 489-517.
9. Hutcheon, B. & Yarom, Y. (2000). Resonance, oscillations and the intrinsic frequency preferences of neurons. *Trends Neurosciences, 23*, 216-222.
10. Itti, L. & Koch, C. (2000). A saliency-based search mechanism for overt and covert shifts of visual attention. *Vision Research, 40*, 1489-1506.
11. Izhikevich, E. (2001). Resonate-and-fire neurons. *Neur. Netw., 14*, 883-894.
12. Kazanovich, Y. & Borisyuk, R.M. (1999). Dynamics of neural networks with a central element. *Neural Networks, 12*, 441-454.
13. Kazanovich, Y. & Borisyuk, R.M. (2002). Object selection by an oscillatory neural network. *BioSystems, 67*, 103-111.
14. Knight, R.T. (1996). Contribution of human hippocampal region to novelty detection. *Nature, 383*, 256-259.
15. Niebur, E. (2002). Electrophysiological correlates of synchronous neural activity and attention: a short review. *BioSystems, 67*, 157-166.
16. Niebur, E. and Koch, C. (1994). A model for the neuronal implementation of selective visual attention based on temporal correlation among neurons. *Journal of Computational Neuroscience, 1*, 141-158.
17. Nishii, J. (1999). Learning model for coupled neural oscillators. *Network, 10*, 213-226.
18. Parra, L., Deco, G., & Miesbach, S. (1996). Statistical independence and novelty detection with information preserving nonlinear maps. *Neural Computation, 8*, 260-269.
19. Sokolov, E.N. (1975). The neuronal mechanism of the orienting reflex. In E.N Sokolov and O.S Vinogradova (Eds.), Neuronal Mechanisms of the Orienting Reflex (pp. 17-35), Erlbaum, Hillsdale, N.J.
20. Torras, C. (1986) Neural network model with rhythm assimilation capacity. *IEEE Trans. Systems, Man, and Cybernetics*, SMC-16, 680-693.
21. Tsotsos, J.K., Culhane, S.M., Wai, W.Y.K., Lai, Y., Davis, N., & Nufl, F. (1995). Modeling visual attention via selective tuning. *Artificial Intelligence, 78*, 507-545.
22. Verschure P. & König, P (1999). On the role of biophysical properties of cortical neurons in binding and segmentation of visual scenes. *Neural Computation, 11*, 1113-1138.
23. Vinogradova, O.S. (1995). Expression, control, and probable functional significance of the neuronal theta-rhythm. Progr. Neurobiol., 45, 523-82.
24. Wang, D.L. (1999). Object selection based on oscillatory correlation. *Neural Networks, 12*, 579-592.
25. Wang, D-L. & Liu, X. (2002). Scene analysis by integrating primitive segmentation and associative memory. *IEEE Trans. Systems, Man, and Cybernetics - Part B: Cybernetics, 32*, 254-268.
26. Wang, D-L. & Terman, D. (1995). Locally excitatory globally inhibitory oscillator network. *IEEE Trans. Neural Networks, 6*, 283-286.

UNDERSTANDING OBJECT FEATURE BINDING THROUGH EXPERIMENTATION AS A PRECURSOR TO MODELLING

C. MAIR, M. SHEPPERD, M. CARTWRIGHT, C. KIRSOPP,
R. PREMRAJ AND D. HEATHCOTE

ESERG, School of Design, Engineering and Computing,
Bournemouth University, UK
E-mail: cmair@bournemouth.ac.uk

In order to explore underlying brain mechanisms and to further understand how and where object feature binding occurs, psychophysical data are analysed and will be modelled using an attractor network. This paper describes psychophysical work and an outline of the proposed model. A rapid serial visual processing paradigm with a post-cue response task was used in three experimental conditions: spatial, temporal and spatio-temporal. Using a 'staircase' procedure, stimulus onset asynchrony for each observer for each condition was set in practice trails to achieve 50% error rates. Results indicate that spatial location information helps bind objects features and temporal location information hinders it. Our expectation is that the proposed neural model will demonstrate a binding mechanism by exhibiting regions of enhanced activity in the location of the target when presented with a partial post-cue. In future work, the model could be lesioned so that neuropsychological phenomena might be exhibited. In such a way, the mechanisms underlying object feature binding might be clarified.

1. Introduction

Working memory (WM) can be described as a system for temporarily holding and manipulating information that is required to perform cognitive tasks (Baddeley[1]). Prefrontal cortex (PFC) plays an important role in WM. Neurons in PFC exhibit sustained activity in the period between presentation of a stimulus and a delayed response (Funahashi *et al.*[5]). Ungerleider, Courtney *et al.*[21] claim that PFC neurons reactivate representations in posterior visual areas via feedback projections, and integrate objects through multiple reverbatory neural assemblies (Raffone and Wolters[10]). Visual information is stored in visual working memory (VWM). Vogel *et al.*[22] suggested that VWM stores integrated objects rather than features. However, Saiki[15] found that such object-based integration occurs only for static images, and as velocity increases, feature detection performance decreases. Reinitz *et al.*[11] and Treisman *et al.*[18] argue that because observers make feature conjunction errors in recall, object features including location are independently processed by memory where they can remain independent. However, despite poor recall, responses are typically better than chance. This suggests the existence of a mechanism by which observers remember feature conjunctions. Examples of proposed integration mechanisms are location-based (Treisman and Gelade[20]; Nissen[8]; Keele *et al.*[7]; Cave and Zimmerman[3]), temporal (von der Malsburg[23]), and verbal (Stefurak and Boynton[16]; Johnson *et al.*[6]). Perception of incorrect combinations of registered features is known as illusory conjunction (IC). ICs are likely to occur from a failure to focus attention (Treisman[19]), when multiple objects are present, and with proximity (Cohen and Ivry[4]). In support of location-based binding, Prinzmetal *et al.*[9] found that ICs are less likely to occur when attention is directed by a cue to a point in space. In most circumstances, feature binding is not a problem for the nervous system, yet it remains an intriguing conceptual problem for neuroscientists. The present study attempts to lead to a further understanding of object feature binding by addressing some of the issues outlined above and testing the following hypotheses. If binding is driven by spatial location (Treisman and Gelade[20]; Cave and Zimmerman[3]), fewer errors should be recorded when stimuli are presented at different spatial locations, than when presented at the same spatial location. Objects appearing spatially closer to the target should be erroneously reported as the target more frequently than objects appearing more distant from the target. In addition, objects appearing temporally closer to the target should be reported as the target in error more frequently than objects appearing more distant in time from the target. If ICs are more likely to occur when multiple objects are present, and with proximity (Cohen and Ivry[4]) there should be a greater occurrence of ICs than chance in all conditions, and more adjacent than distant ICs.

2. Experiment

A within subjects post-cue response task was designed using a battery of sixty-four objects comprising three features in three conditions.

The *stimuli* were 64 rectangular objects with unique feature (colour, size and orientation) combinations. Each feature had four variations: for colour, blue, purple, green and yellow; for size, small to large; and for orientation, horizontal, vertical, left and right diagonal. The stimuli were displayed on a black background. The monitor (17-inch CRT, Dell) had a refresh rate of 75Hz. The experiment was controlled by Macromedia Director 7 software running on a PC. Windows in the room were covered by closed venetian blinds and the room was lit by fluorescent lighting. Observers were positioned such that their faces were 15 inches from the monitor screen and the height of the chair adjusted so that the focal point on the display was the centre of their visual field.

Design and Procedure: Fifteen volunteer adults with normal or corrected vision took part. The three experimental conditions were spatial (SC) (simultaneous at four fixed locations), temporal (TC) (sequential at focal point), and spatio-temporal (STC) (sequential at four fixed locations). Each condition comprised 64 presentations, in each of which a unique set of four objects was pseudo-randomly selected from the battery of 64 objects, to ensure each feature had an equal probability of selection. Within each display set of 4, each object contained each feature once and only once. In SC, the 4 objects were presented simultaneously, each in 1 of 4 positions on the circumference of an imaginary circle (Fig.1) preceded and followed by a checkerboard mask. In TC, the 4 objects were presented sequentially in the focal position separated by a checkerboard mask. In STC, the 4 objects were presented sequentially in the same positions as in SC, separated by the checkerboard mask. When each set of 4 objects had been displayed, the post cue: a coloured 'cloud shape' was presented, and the observers were asked to report the size and orientation of the perceived target (Fig.3). In STC, 2 datasets were collected: one for spatial responses ($STCs$) and one for temporal responses ($STCt$). In SC and $STCs$, a response corresponding to a feature of the object horizontal to the target was recorded as adjacent (A); a response corresponding to a feature of the object vertical to the target was recorded as lateral (L); and a response corresponding to a feature of the object diagonal to the target was recorded as distant (D) (Fig.1). In TC and $STCt$, a response corresponding to a feature of the object $+1$ or -1 to the target was recorded as adjacent (A); a response corresponding to a feature of the object $+2$ or -2 to the target was recorded as intermediate ($D1$); and a response corresponding to a feature of the object $+3$ or -3 to the target was recorded as distant ($D2$) (Fig.2).

Observers were instructed to focus on the centre of the screen indicated by a white dot. Response time was not recorded. Observers carried out 64 trials in each condition to set the error level at 50% for each feature. In order to avoid floor or ceiling effects, the 'staircase procedure' was used to determine the SOA independently for each condition for each observer.

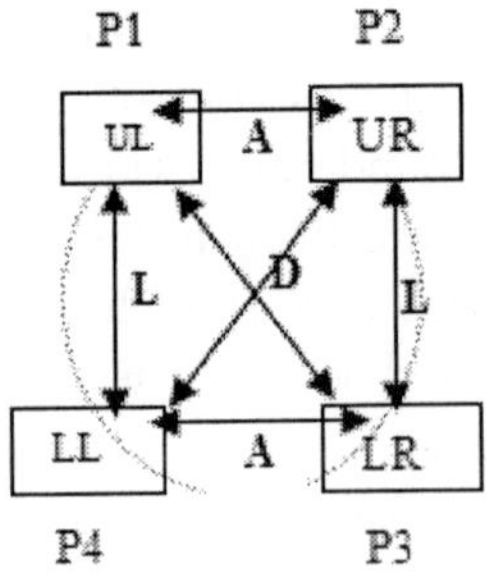

Figure 1. Error coding for spatial (SC and STC) display. P1=position1, P2=position2, P3=position3, P4=position4. UL=upper left, UR=upper right, LR=lower right, LL=lower left. A=adjacent, L=lateral, D=distant.

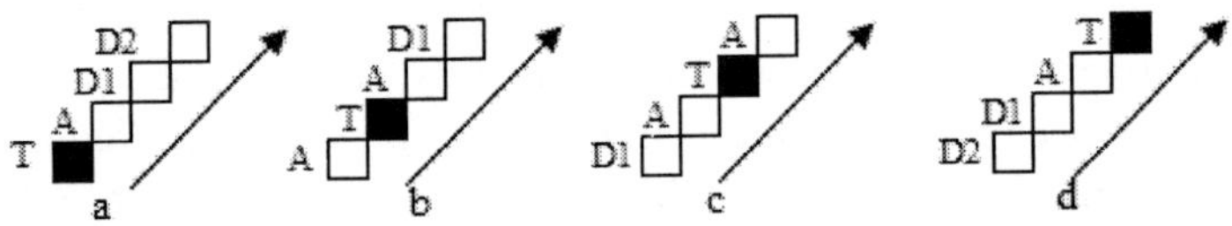

Figure 2. Error coding temporal (TC) display. Target shown as solid. Arrows indicate sequence. T=target, A=adjacent, D1=intermediate, D2=distant

3. Results

Recordings were collected in as 4 datasets: SC, TC and STC space ($STCs$) and STC time ($STCt$). Statistical analyses were carried out using t-tests to make pair-wise comparisons of error position (e.g. spatial errors in SC and $STCs$, A/D, A/L and L/D, and temporal errors in TC and $STCt$, $A/D1$, $A/D2$ and $D1/D2$) and between conditions for specific error positions (e.g. between SC and $STCs$ A/A, D/D and L/L, and between TC and $STCt$ A/A, $D1/D1$ and $D2/D2$). Thus 18 t-tests were conducted. Using the Bonferroni adjustment, alpha was reduced from $p = 0.05$ to $p = 0.003$.

3.1. *SOA*

Mean SOA for SC was 139ms, for TC 172ms, and for STC 137ms. This resulted in mean presentation duration of 417ms in SC, 1548 in TC, and 1233 in STC).

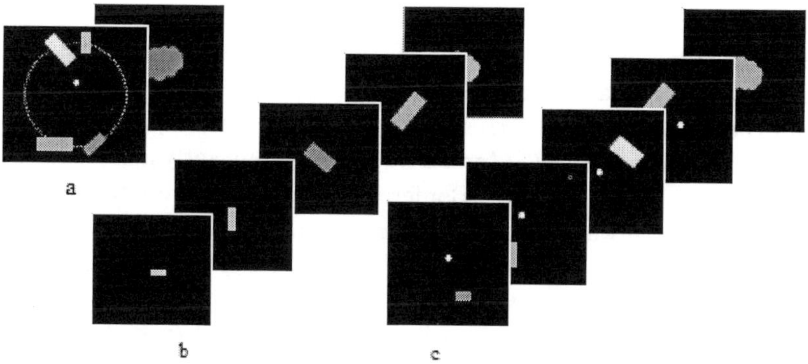

Figure 3. a) Spatial display in which the correct response is the object in position 3 (lower right (LR)); b) Temporal display in which the correct response is the object in position 2; c) Spatio-temporal display in which the correct response is the object in temporal position 2, spatial position 4 (lower left (LL)) . Mask not shown. Circumference shown for illustration only.

3.2. *Spatial and temporal errors*

A total of 7680 responses were recorded (4 datasets: SC spatial, TC temporal, STC spatio-temporal-space and STC spatio-temporal-time, 64 presentations, 2 features each, 15 observers). Despite setting SOA to achieve 50% error rate for each error type, there were significantly more temporal than spatial errors ($p < 0.001$). In addition, there were significantly more errors in $STCs$ than SC ($p < 0.001$). There were also more errors in TC than $STCt$, but using the Bonferroni adjustment, this difference was not significant ($p = 0.005$). Observers had an equal chance (33%) of making each type of spatial error, but a 50% chance of making temporally adjacent error, 33.3% chance of a temporally intermediate error, and 16.7% chance of a temporally distant error. The greatest number of errors (1028) were recorded in TC (7% more than expected), the fewest (734) in SC (24% fewer than expected), 953 errors were recorded in STC (1% less than expected).

3.3. *Adjacent and distant errors*

'Expected' values were determined by setting the SOA to achieve a 50% error rate for each observer in each condition. Overall, there were more adjacent (1673) errors and more distant (773) errors ($p = 0.004$ and $p = 0.008$).

In SC, lateral and temporally intermediate errors (1146) were not significantly greater than expected. In $STCs$, that is the recording of spatial errors in STC, actual adjacent errors were significantly greater ($p < 0.001$) than expected, but distant errors were not ($p = 0.005$). Spatially adjacent errors between conditions (SC and STC) were significantly greater in $STCs$ than SC ($p < 0.001$), but after Bonferroni adjustment there was no significant difference between spatially distant errors ($p = 0.02$). Differences between conditions for temporal errors (TC and $STCt$) were not significantly different, but in both conditions, temporal errors of all types were greater than expected: adjacent errors in both TC and $STCt$ ($p < 0.001$); intermediate errors in TC ($p = 0.004$), intermediate errors in $STCt$ ($p < 0.001$); distant errors in TC ($p = 0.012$) and in $STCt$ ($p = 0.006$).

3.4. *Error effects*

Table 1. One-tailed t-tests for effect of spatial and temporal positional error.

	Spatial			*Temporal*	
	$A - L$	$L - D$	$A - D$		$A - (D1 + D2)$
SC	35	35	0	*TC*	52
STC	142	90	−52	*STC*	52
Temporal Effect	107	55	52	*Spatial Effect*	0

To recap, spatial errors were recorded in SC and in $STCs$, and temporal errors recorded in TC and $STCt$. The spatial positional error effect is defined as the difference between $A - L$, $A - D$, and $L - D$ errors. The temporal positional effect is defined as $A - (D1 + D2)$. The spatial error effect (between SC and $STCs$) is a result of temporal properties; the temporal error effect (between TC and $STCt$) is a result of spatial properties (Table 1).

Table 2. One-tailed t-test results for significance between effect of positional error.

Spatial			*Temporal*		
Position	*SC*	*STCSpace*	*Position*	*ST*	*STCTime*
$A - L/A - D$	0.5	0.016	$A - D1/A - D2$	> 0.001	> 0.001
$A - L/L - D$	0.163	> 0.001	$A - D1/D1 - D2$	0.182	0.292
$A - D/L - D$	0.067	> 0.001	$A - D2/D1 - D2$	> 0.001	> 0.001

Error effects in $STCs$ were highly significant. Temporal error effects in both TC and $STCt$ were highly significant at $A - D1/A - D2$ and $A - D2/D1 - D2$ (Table 2). To compare spatial and temporal error positional effects, comparisons were made between SC and $STCs$, and TC and $STCt$. Spatial error effects were significant only for $P1 - P2$, while temporal error effects were insignificant (Table 3).

Table 3. One-tailed t-tests for effect of spatial and temporal positional error.

Position	$SC - STCs$	*Position*	$TC - STCt$
$A - L$	0.004	$A - D1$	0.458
$A - D$	0.158	$A - D2$	0.067
$L - D$	0.1	$D1 - D2$	0.225

3.5. *Illusory Conjunctions (ICs)*

An IC was recorded when erroneous responses to both features corresponded to the colour of a non-target. Occurrence of ICs (605) was not significantly greatly than chance. Overall most were adjacent (38% of all adjacent errors), but these were recorded mainly in temporal conditions (41% of all adjacent error in TC, and 45% in $STCt$). Fewest distant ICs were recorded in temporal conditions (14% in TC, and 23% in $STCt$). Spatial ICs were more closely distributed across error types.

4. Discussion

Displays in TC and STC had longer durations than SC. To remove effects of primacy and recency on recall, SOA was set a priori to achieve 50% in each condition. More errors were made when objects were presented at the same spatial location (TC) than when they were presented at individual locations (SC and STC). When objects were presented at different locations, but were separated in time (STC), there were fewer temporal errors than in TC, and more spatial errors than in SC. This implies that separating objects in space aids integration in recall, whereas separating objects in time is detrimental. Most errors were adjacent in both space and time, but spatially adjacent errors were highly significantly greater when objects were presented sequentially (STC) than when presented simultaneously (SC). The spatial error effect was more than three times greater than the temporal error effect. The temporal positional error effect ($A - (D1 + D2)$) was equal, whereas the spatial positional error effect was insignificant between positions in *SC*, but significant for all positions in *STC*. Results from this study suggest location information is important for binding, but it is most important when objects are not separated in time. Indeed, Keele *et al.*[7] claimed that although location is primary, time of stimulus occurrence must be taken into account. Saiki[15] claims that location-based integration applies only to static objects. In *TC* and *STC* stimuli were presented sequentially with brief exposure. The stimuli might have been perceived as moving and therefore not correctly retained as integrated objects in memory.

According to earlier research (e.g. Cohen and Ivry[4]) ICs occur mostly when multiple objects are present, and with proximity. However, ICs were not significantly greater than chance in any condition, but more ICs occurred mostly in *TC* and *STCt*, and more were adjacent than distant. To more fully explore the data from this experiment and ultimately to further understand the mechanisms of object feature binding, a neural model is under development. The following sections describe the early stages of development and potential future modifications of the proposed model.

5. The neural model under development

The aim of the proposed neural model is to further understand where and how feature integration takes place. Prefrontal cortex (PFC) has been found to exhibit sustained activity during recall, even in the absence of a stimulus (Renart *et al.*[12]), when potential match stimuli are shown, possibly through modifiable recurrent collateral connections between neighbouring pyramidal cells in an autoassociative neural network. The neural model will be a representation of a network of PFC pyramidal cells. Membrane potential is reset after each spike, therefore in order to maintain activity in the absence of a stimulus, real neurons feedback on themselves. This can be represented in an autoassociative network model with collateral recurrent connections tuned to avoid the possibility of each node becoming active once a subset is activated. In addition, to enhance biological plausibility, the nodes should represent integrate and fire (IF) neurons

The model is defined following Trappenberg[17], Rolls and Deco[13] and Rolls and Treves[14]. Nodes have diluted connectivity to give a non-symmetric synaptic matrix generated by Hebbian learning on random patterns, and the activation function is threshold linear to enable continuously variable firing rates. Weights are initialised to random values before learning starts. Inputs are mainly excitatory and mutual inhibition is implemented by inhibitory interneurons which comprise 10% of the 10,000 network nodes. The basic architecture is a one-layer (2-D) network of $N * N$ nodes with a set of inputs that make modifiable excitatory synapses with the output neurons. The output cells compete by means of the softmax function (Bridle[2]), and at each time step, all nodes are simultaneously updated by random sequence. Spatial location is an important factor in the current study and is represented in the model as 'the centre of mass of an activity packet' (Trappenberg[17] p.209). The post-cue stimulus comprises one feature (colour) of one of the four composite objects in the original input. When applied, the post-cue activates the nodes with a factor proportional to a Gaussian centred on that position and decaying as the cue gets further from that centre. In doing so, the composite object originally displayed at the spatial location, is recalled from the partial cue. This is feasible in conditions where stimuli occupy unique spatial locations. However, when objects are presented sequentially at the same spatial position, as in the temporal position in the current study, unique spatial information is not available. We suggest that in such conditions, memories 'drift' (Trappenberg[17]) and recall is more prone to error than in situations in which each memory has a unique spatial location. We propose that when the next object is presented, firing activity for each object, represented graphically as a 'bump', will subside and 'drift' to a neighbouring location rather than disappear. A new 'bump' will emerge at the spatial location of the first, and so on. At recall, competition takes place between the stored memories to result in enhanced activity of the memory with the feature corresponding to the cue, and suppressed firing of the other memories. In such a way, temporal information, which has no unique spatial location, might be endowed with spatial properties in memory to facilitate recall.

6. Conclusion

Psychophysical results from this and other studies have shown spatial information helps bind object features. By analysing output from the proposed neural model it might be possible to determine the dynamics of spatial and temporal errors, particularly in the spatio-temporal condition when objects have neither a unique spatial nor temporal location and it was shown that correct recall is impoverished by temporal properties and enhanced by spatial ones. In addition, understanding how memories develop into stable states, might lead to a better understanding of how temporal location properties are stored and whether they are transferred to spatial location properties in recall. Further analyses of the psychophysical data and the model are planned. For example, because of its architecture, output from the basic model will demonstrate rapid and correct recall. However, humans make binding errors, and responses from observers in the experiment described earlier were tuned to achieve a 50% error rate. Therefore, the neural model will have to be adapted in order that it too achieves a 50% error rate. Further analyses should determine whether a particular feature (orientation or size) is dominant in binding, and if so, how do the dimensions of that feature impact on recall? What is the effect of the post-cue presentation on target recall in the following presentation? Futhermore, an experiment, incorporating a simple verbal activity could be conducted so that verbalization, which in some cases has been shown to enhance recall (Stefurak and Boynton[16]; Johnson *et al.*[6]) of object features would not be possible. Despite continued effort in the research community and a large literature on binding, the mechanisms that allow individual features to be bound rapidly and, for the most part, correctly into composite objects remain poorly understood. The aim of the work in hand is to elucidate on these mechanisms.

References

1. A. D. Baddeley, *Working Memory.* Oxford University Press (1986)
2. J. S. Bridle, Probabilistic interpretation of feedforward classification network outputs, with relationships to statistical pattern recognition. *Neurocomputing: Algorithms, Architectures and Applications*, F. Fogelman Soulie and J. Herault (Eds.) Springer Verlag (1990)
3. K. R. Cave and J. M. Zimmerman, Flexibility in spatial attention before and after practice. *Psychological Science* **8**: 399-403 (1997).
4. A. Cohen and R. Ivry, Density effects in conjunction search: evidence for a coarse location mechanism of feature integration. *Journal of Experimental Psychology: Human Perception and Performance* **17**: 891-901 (1991)
5. S. Funahashi, C. J. Bruce and P. S. Goldman-Rakic, Visuospatial coding in primate prefrontal neurons revealed by oculomotor paradigms. *Journal of Neurophysiology* **63**: 814 (1990)
6. T. R. W. Johnson, W. Hongbin, H. Wang and J. Zhang, A model of spatio-temporal coding of memory for multidimensional stimuli. *Proceedings of the 24th Annual Conference of the Cognitive Science Society* Lawrence Erlbaum (2002)
7. S. W. Keele, A. Cohen, R. Ivry, M. Liotti and P. Yee, Tests of a temporal theory of attentional binding. *Journal of Experimental Psychology: Human Perception and Performance* **14**(3): 444-452 (1988)
8. M. J. Nissen, Accessing features and objects. *Attention and Performance*

XI, M. I. Posner and O. S. M. Marin. Hillsdale (Eds.), Lawrence Erlbaum Associates (1985)

9. W. Prinzmetal, D. E. Presti and M. I. Posner, Does attention affect visual feature integration? *Journal of Experimental Psychology: Human Perception and Performance* **12**: 361-369 (1986)
10. A. Raffone and G. Wolters, A cortical mechanism for binding in visual working memory. *Journal of Cognitive Neuroscience* **13**: 766-785 (2001)
11. M. T. Reinitz, W. J. Lammers and B. P. Cochran, Memory conjunction errors. *Memory and Cognition* **20**(1): 1-11 (1985)
12. A. Renart, N. Parga and E. T. Rolls, A recurrent model of the interaction between the prefrontal cortex and inferior temporal cortex in delay memory tasks. *Advances in Neural Information Processing Systems.* S. Solla, T. Leen and K. R. Mueller (Eds). MIT Press. **12**: 171-177 (2000)
13. E. T. Rolls and G. Deco, *Computational Neuroscience of Vision.* Oxford University Press (2002)
14. E. T. Rolls and A. Treves, *Neural Networks and Brain Function*, Oxford University Press (1998)
15. J. Saiki, Spatiotemporal characteristics of dynamic feature binding in visual working memory. *Vision Research* **43**: 2107-2123 (2003)
16. D. I. Stefurak and R. M. Boynton, Independence of memory for categorically different colours and shapes. *Perception and Psychophys.* **39**(3): 164-174 (1986)
17. T. P. Trappenberg, *Fundamentals of Computational Neuroscience.* Oxford University Press (2002)
18. A. Treisman, Focussed attention in the perception and retrieval of multidimensional stimuli. *Perception and Psychophysics*, **22**: 1-11 (1977)
19. A. Treisman, The binding problem. *Current Opinion in Neurobiology.* **6**(2): 171-178 (1996)
20. A. Treisman and G. Gelade, A feature-integration theory of attention. *Cognitive Psychology* **12**(1): 97-136 (1980)
21. L. G. Ungerleider, S. M. Courtney and J. V. Haxby, A neural system for human visual working memory. *Proceedings of the National Academy of Science* USA. **95**(3): 883-890 (1998)
22. E. K. Vogel, G. F. Goodman and S. J. Luck, Storage of features, conjunctions, and objects in visual working memory. *Journal of Experimental Psychology: Human Perception and Performance* **27**(1): 92-114 (2001)
23. C. Von der Malsburg, The Correlation Theory of Brain Function. *Biophysical Chemistry* **MPI**: 81-82 (1981)